ALESSANDRO VOLTA

# RELENTLESS VISIONARY

**Michael Berick**

THE
MENTORIS
PROJECT

The author has made every effort to ensure that the accuracy of the information within this book was correct at time of publication. The author does not assume and hereby disclaims any liability to any party for any loss, damage, or disruption caused by errors or omissions, whether such errors or omissions result from accident, negligence, or any other cause.

Barbera Foundation, Inc.
P.O. Box 1019
Temple City, CA 91780

Copyright © 2019 Barbera Foundation, Inc.
Cover photo: World History Archive / Alamy Stock Photo
Cover design: Suzanne Turpin

More information at www.mentorisproject.org

ISBN: 978-1-947431-30-0

Library of Congress Control Number: 2020900491

All net proceeds from the sale of this book will be donated to Barbera Foundation, Inc. whose mission is to support educational initiatives that foster an appreciation of history and culture to encourage and inspire young people to create a stronger future.

The Mentoris Project is a series of novels and biographies about the lives of great men and women who have changed history through their contributions as scientists, inventors, explorers, thinkers, and creators. The Barbera Foundation sponsors this series in the hope that, like a mentor, each book will inspire the reader to discover how she or he can make a positive contribution to society.

# Contents

# Foreword

First and foremost, Mentor was a person. We tend to think of the word *mentor* as a noun (a mentor) or a verb (to mentor), but there is a very human dimension embedded in the term. Mentor appears in Homer's *Odyssey* as the old friend entrusted to care for Odysseus's household and his son Telemachus during the Trojan War. When years pass and Telemachus sets out to search for his missing father, the goddess Athena assumes the form of Mentor to accompany him. The human being welcomes a human form for counsel. From its very origins, becoming a mentor is a transcendent act; it carries with it something of the holy.

The Mentoris Project sets out on an Athena-like mission: We hope the books that form this series will be an inspiration to all those who are seekers, to those of the twenty-first century who are on their own odysseys, trying to find enduring principles that will guide them to a spiritual home. The stories that comprise the series are all deeply human. These books dramatize the lives of great men and women whose stories bridge the ancient and the modern, taking many forms, just as Athena did, but always holding up a light for those living today.

Whether in novel form or traditional biography, these books plumb the individual characters of our heroes' journeys.

The power of storytelling has always been to envelop the reader in a vivid and continuous dream, and to forge a link with the subject. Our goal is for that link to guide the reader home with a new inspiration.

What is a mentor? A guide, a moral compass, an inspiration. A friend who points you toward true north. We hope that the Mentoris Project will become that friend, and it will help us all transcend our daily lives with something that can only be called holy.

—Robert J. Barbera, President, Barbera Foundation
—Ken LaZebnik, Founding Editor, The Mentoris Project

# Preface

F rogs' legs, the torpedo fish and the cell phone are all connected. In fact, the cell phone owes its existence to frog legs and the torpedo fish. The same can be said for a range of everyday essentials, including computers, cars, toys, and tools.

What these devices have in common is that they each operate with a battery. The link between batteries and frog legs and the torpedo fish will be revealed in the following pages, but the lynchpin is a man named Alessandro Volta.

Alessandro Giuseppe Antonio Anastasio Volta was an Italian scientist who was born in Como, Italy in 1745 and died there in 1827. In his life he traveled throughout Europe and became a renowned electricity experimenter (as scientists were often called at that time), credited with many significant discoveries and inventions. He was the first person to identify the gas known as methane, and created the first authoritative list of conducting metals. He studied atmospheric science and made advancements in the field of meteorology. He also created an array of instruments, including the "electrical pistol." His most important invention, and the crowning glory of his career, was the Voltaic pile, which is now recognized as the first electric battery. Electricity is the primary form of energy that powers our world, and batteries of all sizes and levels of power perform incalculable functions on a daily basis.

Although his name may not be as well known as Thomas Alva Edison or Nikolas Tesla, Alessandro Volta was an influential scientist of his generation. Volta's work and contributions continue to affect our lives in ways big and small, obvious and not-so-obvious. For example, a connection can be drawn between his work and Mary Shelley's novel *Frankenstein*.

Alessandro Volta was an extraordinary scientist, and lived an extraordinary life. He was born and raised in a small Italian town, and it is said that he didn't speak for the first several years of his life. His family lived far from Europe's intellectual centers, and he had no formal education, but this did not impede Volta's intelligence or curiosity.

A product of the Age of Enlightenment—a time when ideas on reason, science, literature, and liberty took center stage—Volta employed a very modern, hands-on approach to his work. He built his career by seeking out influential people and key opportunities that would help him achieve his goals. He met Benjamin Franklin and other great men in his field, and he socialized with the rich and with royalty. In fact, Napoleon Bonaparte was a fan and became an important patron. Volta's success is all the more intriguing because of the many challenges he had to overcome.

Alessandro Volta saw things not just as they were, but as what they could be. He was a disrupter, an innovator and a visionary.

Above all, however, he was relentless—Volta's hunger to create, and his drive to invent and discover were simply remarkable.

*Relentless Visionary* covers the journey of Volta's life, revealing his steps to becoming a celebrated scientist and compelling historical figure, and examines his legacy within the context of seventeenth- and eighteenth-century history. Readers will learn how Volta changed his world and why he remains so relevant today.

# Chapter One

## ROOTS AND ORIGINS

### VOLTA'S FAMILY

The world was rapidly changing in 1745. The Age of Enlightenment, a phase of European history that championed human reason, was flourishing. The Scientific Revolution was ongoing, and the Industrial Revolution was not far off. Much of Europe was in a state of turmoil. Nearly every European power was entrenched in the War of the Austrian Succession, and in addition, the Scottish revolt against the British, known as the Jacobite Rising, was just beginning.

Political events, however, were of little, if any, concern in the Volta household. The Voltas were more interested in the arrival of the latest member to their family, a baby boy they named Alessandro Giuseppe Antonio Anastasio Volta. Born on February 18, 1745, Alessandro was the youngest son of Filippo and Maddalena Inzaghi Volta, and was born into a family with four sisters and four brothers.

The Voltas lived in the town of Como in the Lombardy region of what is now Italy. Located about twenty-five miles north of Milan, Como rests near the Alps on the southwestern shore of the lake that bears its name. When Alessandro was born, Como was a picturesque outpost of the Austrian Empire. The town, and the rest of Lombardy, had come under Austrian rule in 1714 during the Hapsburg regime. Except for a brief time of French rule during the years of 1796 through 1815, the region remained under Austrian control. The city was a trade center with a large silk manufacturing industry that attracted salesmen from around Europe. This international activity, coupled with proximity to both France and Austria, would influence Volta's life, particularly as a young man.

Filippo Volta had spent eleven years in the Jesuit order by the time he married Maddalena Inzaghi. He was forty-one years old at that time; she was twenty-two years his junior. Volta's parents came from families of lesser nobility. Filippo's forefathers included Martino Volta, a successful Venetian wool merchant during the age of Christopher Columbus, while Maddalena came from a noble family in Graz, a city in modern-day Austria. Though they each enjoyed a certain degree of prestige and privilege in their lives, they were not wealthy.

Not much is known about Filippo, but the little that is known isn't particularly complimentary. Described as an unreliable, possibly troubled individual, Filippo earned a reputation for his lavish spending and not his business acumen. A well-known example took place at Como's 1750 carnival celebration, where Filippo organized two grand dinners for his friends and associates, even though his own large family of nine children was struggling financially. Late in his life, Volta wrote that his family had been left with a 17,000 lire debt when his father died while Alessandro was a young child.

Following Filippo's death, Maddalena took Alessandro and two of his sisters, Chiara and Marianna, to live with Alessandro's uncle and namesake, Alessandro Volta. The elder Alessandro served as the Como cathedral's archdeacon and played a stepfather-like role in young Alessandro's life.

Volta didn't have a close relationship with either of his parents. His estrangement from his father is easy to understand given Filippo's irresponsible behavior, particularly in financial matters. That Volta was not close to his mother is more curious. When his mother was on her deathbed in 1782, Alessandro was in Milan, just twenty-five miles from Como. Instead of going home to be with his mother, Alessandro instead chose to stay in Milan and keep track of her health from there. She died several days later while he was still in Milan.

The Volta family's financial situation improved in 1756 when the Volta sons, principally Alessandro and his brother Luigi, received an inheritance from their wealthiest relative, great-uncle Nicolò Stampa, who bequeathed them a substantial revenue-producing trust. Luigi, who at age fifteen was four years older than Alessandro, was appointed the main heir while Alessandro was named in a secondary position. The two brothers consequently occupied an important role in their family. As beneficiaries of the trust they provided their family's main income source.

The Volta brothers would spend a good part of the year, from springtime to early winter, touring the family properties—approximately nine separate small estates and houses located across the regions around Como and Milan. They enjoyed the rental money that these properties provided and a certain amount of leisure that came from the inheritance and the rental revenue.

While the Voltas were more comfortable due to the inheritance, they were not wealthy. In fact, Luigi remained the trust's primary heir, holding the reins of the family's financial affairs until Alessandro was well into middle age. It was only when Alessandro married in 1794, at age forty-nine, that Luigi ceded him fifty percent of the rents from the Stampa properties.

## THE EARLY EDUCATION OF A GENIUS

As a young child, Alessandro Volta showed no signs of becoming a brilliant scientist. In fact, his family feared that he might be mentally challenged because he did not speak a word in the first few years of his life, and appeared to be dim-witted. It was only after he turned four that Alessandro started to talk; his first word was "no!" Perhaps this was a sign of his against-the-grain nature that blossomed as he grew older.

Once young Alessandro found speech, his development accelerated. By the time he was seven, he was among the top students in his class. His family was surprised by their son's sudden and rapid intellectual growth spurt. His father said, "We had a jewel in the house, but did not know it."

The delayed developmental that Alessandro experienced is not unusual for youngsters who grow up to be highly intelligent adults. Albert Einstein, for example, did not begin talking until he was around age three, and noted physicists Richard Feynman and Edward Teller were also late talkers. Economist and social critic Thomas Sowell even coined the term "The Einstein Syndrome" for this type of late-talking child.

Not much is known about Volta's early formal education. Some scholars suggest that he was homeschooled as a child, which was not unusual in the eighteenth century. When Alessandro later became the superintendent of Como's public school, he endorsed homeschooling, and thought that it should be supported by the state. Some believe that Alessandro entered a school of rhetoric in Como when he was seven, around the time that his father died in 1752. Wherever Alessandro received his early education, he became a quick and voracious learner. In 1758, at the age of thirteen, he entered Como's Jesuit College, a small but well-established institution that had been founded 200 years earlier. During his years there, young Alessandro received a classic Jesuit education, studying physics, rhetoric, and philosophy.

## A YOUNG SCIENTIST AND A POET

While Alessandro Volta displayed an early curiosity about the sciences, he had many academic interests. He enjoyed the study of the humanities, and demonstrated an aptitude for languages. In addition to his native Italian, he learned Latin, French, and English, and was able to read Dutch and Spanish. Alessandro's fondness for French tragedies and his desire to read French scientific scholarship have been noted as his motivation for learning the French language. His aptitude for linguistics served him throughout his professional life—he could read scientific literature in a variety of languages as well as communicate with scientists around Europe through letters and in person.

Poetry was another interest. Alessandro gravitated to epic poems, especially those written by the German poet Friedrich Gottlieb Klopstock (*der Messias*) and the English poets Edward

Young (*Night Thoughts*) and John Milton, whose *Paradise Lost* he read in a French translation. He enjoyed reading Italian poets Torquato Tasso, Carlo Innocenzo Frugoni, and Gabriello Chiabrera. Alessandro was fond of contemporary and classic oratory; Cicero was one of his favorites.

Writing poetry presented Volta with an opportunity to combine his interests in science and the humanities. When he was about eighteen, the young Volta composed an epic poem— some 500 verses—that addressed contemporary discoveries in natural sciences and philosophies. In this untitled poem, Alessandro sought to rationally explain such things as gunpowder and fireworks, which were then viewed as magical. The poem also touches upon the topic of electricity, and mentions "philosophers" (scientists were considered to be in this group) such as Benjamin Franklin, Giovanni Battista (or Giambattista) Beccaria and Jean-Antoine Nollet, who all explored electricity to some extent and would factor in Alessandro's later career. This poem hints that Alessandro performed experiments to provide evidence of his scientific theories. He planned to devote a specific poem to the topic of electricity, although no such poem, if he ever wrote it, survives today.

Writing poetry remained a lifelong passion. Alessandro described himself in a 1768 letter to the Swiss poet Salomon Gessner, as being "fond of everything poetic." To commemorate his friend and fellow electrical experimenter H. B. de Saussure's successful summit of Mont Blanc in the Alps, Alessandro penned a poem in 1787 comprising 199 verses.

The pursuit of knowledge and intellectual inquiry, especially in natural philosophy—the study of natural and physical sciences—were keen interests of the teenage Alessandro Volta. During this period, he generated controversy in Como by arguing that animals, like humans, had souls. It was a position that ran against the prevailing opinion in his hometown. Among those who vigorously disagreed with him was his good friend Giulio Cesar Gattoni. While he was interested in natural philosophy, Gattoni was also a staunch Catholic—his religious beliefs conflicted with Alessandro's ideas.

Alessandro's relatives removed him from the Jesuit College after just a year or two, even though the school offered free admission. His family worried about the political troubles that the Jesuits were encountering in some European states. They also believed that Alessandro's philosophy professor had attempted to persuade him to join the Jesuits and become a priest. The possibility of Alessandro being a priest troubled his family because the 1756 Stampa trust stipulated that the substantial inheritance from that the deed would go to a male descendent, and this would not happen if he were to become a priest. Since Alessandro's brothers were already in the church, the family needed Alessandro to marry and have children to preserve the deed of the trust. It was understandable why the Voltas would steer their sole remaining male heir away from priesthood, despite the fact that church was a strong presence and a source of income in their family (two of Alessandro's sisters became nuns.)

Alessandro's family, and particularly his uncle, pushed him to study law and to get a financially stable job such as that of a magistrate. But this plan held no interest to Alessandro. He made it clear that he wanted to study philosophy. Eventually his family acquiesced and allowed him to pursue this area of study. His mother did not play a crucial role in the discussions about the career of sixteen-year-old Alessandro.

And so Alessandro Volta began to study philosophy rather than law or religion. He enrolled in a local seminary—most historians believe it was either Santa Caterina or Seminary Benzi—whose curriculum of philosophy, grammar and rhetoric did not differ markedly from that of the Jesuit College. Significantly, however, this new school included many students like Alessandro, who intended to have lay careers unrelated to the church.

Alessandro was an avid, intense learner. His thirst for knowledge opened his mind to a broad spectrum of topics, including literature and languages and a variety of sciences. His interest and curiosity were driven by a sense of imagination and creativity.

He possessed an especially strong ability to concentrate, and he was known to be so consumed while working on an experiment that he would miss meals and skip sleep. He paid little attention to his clothing. His focus was solely on his studies; he had no interest in fashion or frivolity.

When he was faced with a scientific problem he could not solve, Alessandro doggedly sought out explanations. This was particularly true with problems related to nature. He employed the full force of his inquisitive mind and thorough, hands-on methodology until he reached understanding.

The budding "electrician," as scientists of electricity were often called at the time, read and engaged in ideas of physics, an influence from his earlier Jesuit College studies. In addition to his education in the classical tradition, Alessandro read seventeenth-century authors who reinforced his worldview and provided him with rich inspiration. Isaac Newton, Benjamin Franklin, Jean-Antoine Nollet, Giovanni Beccaria, and Pieter van Musschenbroek were among his favorites.

## YOUNG VOLTA LOOKS BEYOND COMO

Alessandro was driven, both personally and intellectually, to create a life for himself that his provincial hometown could not provide. The first step was to break away from his family's expectations. In his later years, Alessandro reflected that he found their routine life in Como, as comfortable as it was, to be mundane and stifling. It was simply too small. He desired a grander and broader life, and he vowed to escape the isolating existence in Como.

His quest for learning took Volta beyond the walls of his home and school, and of his town. Through his scientific endeavors, as well as his writing interests, he sought to position himself for a cosmopolitan lifestyle. Volta's curiosity about the world outside Como, and even Italy, can be seen in the impressive list of continental writers, philosophers, and scientists he admired in school. As a young man, Volta also caught a glimpse of the world at large by getting to know members of the local business community as well as clergy, academics, aristocrats, and intellectuals around Lombardy.

Volta's broad group of associates contributed to his remarkable professional successes. Whether or not he did it consciously, Volta engaged in a creative cross-pollination that proved to be highly productive. By associating with natural philosophers, expert electricians, amateur electricians, university professors, and medical instrument makers, he created an intellectually stimulating community that could be viewed as a scholarly version of the artistic salons that were popular during the second half of the eighteenth century.

Besides sharing many of his intellectual interests, a number of the men who Volta encountered as he started his professional journey aspired to be natural philosophers just as he did. Among these was his long-time friend Giulio Gattoni. A few years older than Volta, Gattoni came from a wealthy family. Sharing Volta's curiosity in electricity, he would often provide Volta with books, and even let him work on experiments in the physics lab in his own house. Gattoni later received a good deal of recognition in northern Italy for setting up Como's first lightning rod, as well as for his collection of natural and antiquarian objects.

## A MAN WITH A PLAN

While Volta can be seen as a bit of an eighteenth-century rebel, he was not a true revolutionary, but rather a moderate with a rebellious streak. Volta's ambitions might have been restricted by social traditions, and his ideas certainly ran counter to the prevailing views at times. However, his unorthodox mindset did not lead him to form a hostile view of humanity or attempt to topple society. Just the opposite; Volta wanted to improve the world.

Similarly, Alessandro Volta would only go so far in breaking away from his own heritage and its teachings, though he was drawn to other cultures. His childhood upbringing and education grounded him in ways that led him to respect traditions even as he sought to free himself from their more restrictive aspects. In physics, Volta found a secular field of study that embraced the rational. The concept of electricity did not come charged with religious controversy on topics such as the age of earth or whether or not animals had souls like humans—ideas that Volta vigorously debated at one time, but later learned to avoid.

During the mid-seventeenth century, the work of the amateur scientist in cultural circles was celebrated much in the way as that of musicians and writers. Volta, however, was not willing to be viewed as an amateur philosopher; there were plenty of those in Lombardy and all around Italy. He wanted his work as an "electrician" to be taken seriously, and he did not care whether or not it was fashionable. Natural philosophy, and particularly physics, was of great interest and provided him with the perfect outlet for his ambitions and worldview.

Because he came from lesser nobility and had chosen not to go into a career in law or in the church, both of which offered a steady income, Volta needed to achieve a different route to social and cultural standing. He aimed to become known as a serious natural philosopher, which he hoped would then build and spread his reputation beyond Lombardy and the Italian providences to greater Europe.

Volta exhibited impressive drive—and an ability to advance in the path he chose while still a teenager—to become an electrician and natural philosopher. Even without matriculating at a university, he began to formulate theories of electricity, experimenting on his own and promoting his ideas and findings.

In the early 1760s, the main Lombard scholar studying electricity was Carlo Barletti, but this man was just a decade older than Volta and not yet an established electrician. To find someone with the sufficient fame and authority necessary to help advance his career, Volta surveyed the leading electricians and natural philosophers, looking for those whose work he respected.

One such person was Benjamin Franklin; however, the distance between Como and Philadelphia was too far to make regular contact feasible. Volta instead chose a worthy Franklin substitute—Giovanni Beccaria. An acolyte of Franklin's, Beccaria lived only ninety miles from Como. The University of Turin professor of physics also enjoyed high standing among the electricians of Europe and the New World. Moreover, Volta was aware that the Austrian emperor, Joseph II, had visited Beccaria and attended an event in Turin where the distinguished professor demonstrated his electrical experiments.

In the mid-seventeenth century, communication options were limited. The best method was to send a letter, which Volta did to contact Beccaria in 1763. At first, the venerated scientist was a sporadic, and not a particularly well-tempered correspondent. His behavior is understandable considering that Volta was a relatively unknown scientist who was a generation younger than Beccaria.

By 1765, however, Beccaria had begun to reply to Volta regularly. He even sent Volta, in late 1766, the proofs of his forthcoming article for the prestigious journal, the *Philosophical Transactions* of the Royal Society of London. This gesture reveals that Beccaria viewed the much younger Volta as something of a professional equal, underscoring the value of Volta's persistence in writing to him. The significance of Beccaria's correspondence was not forgotten by Volta over the years. When he published his first treatise in 1769, Volta addressed it to Beccaria. He also would choose the same publication, *Philosophical Transactions*, to announce the invention of the Voltaic pile more than three decades later.

While Volta corresponded with Beccaria, he also started writing to Jean-Antoine Nollet, who, like Beccaria, was then a well-established expert in the scientific community. The Paris-based Nollet also had the cachet of coming from that noted world city, which had a reputation as a glamorous center of learning. Moreover, when Volta contracted Nollet, the noted French physicist and electrical experimenter already had a twenty-five-year track record of visiting Italy and engaging with Italy's electricians. In fact, his status had increased in the Italian peninsula during the 1760s because two of his works appeared in Italian translation in 1761 and 1762.

While writing to Nollet, Volta expressed his idea that electrical attraction was similar to Newton's law of gravitational attraction. Nollet responded enthusiastically, stating, "Your letter on the nature of electricity brought me great pleasure. I marvel that you were able to derive from Newtonian attraction the laws of electrical phenomena but I fear a successful outcome to be most difficult. I do not know another physicist with the daring to enter the path; the glory will be all yours if you succeed in accomplishing it laudably."

Nollet's words proved to be quite prophetic. Volta did find great success with his electrical experimentations, although not necessarily with what he wrote about in that particular letter. Over the years, Volta and Nollet continued to exchange letters as well as books. In writing to Volta in September 1767, only a few years before he passed away, Nollet also included the gift of a small book for his faithful correspondent.

The eighteen-year-old Volta demonstrated a youthful audacity by contacting Nollet and Beccaria out of the blue, especially since he lacked any bona fide credentials or even a mutual connection to justify doing so. Volta displayed daring and deference in equal measure by not only writing these recognized authorities, but by discussing his ideas about electricity, magnetism, and related topics. Additionally, he showed persistence in communicating with both men, who had higher priorities than corresponding with this younger man whose letters they undoubtedly regarded—at least initially—with a healthy amount of skepticism. Volta succeeded in becoming a consistent correspondent with both Nollet and Beccaria.

In contacting these men, Volta also revealed a certain political savvy. At the time, Nollet and the Beccaria-Franklin camp were engaged in a high-profile intellectual debate regarding the theory of electricity. By corresponding with both men, Volta consequently was able to share his views with two leading exponents of the two differing schools of electrical thought. Furthermore, he added some minor value to each side, since the two camps sought to gain additional supporters in their intellectual debates.

Volta's correspondence with Nollet and Beccaria, moreover, established a habit that lasted for years to come. Throughout his life, Volta reached out to highly regarded scientists and courted influential people outside the science community to cultivate his professional career.

Volta wished to become a member of the "Republic of Letters," a transnational, interdisciplinary community of intellectual Europeans who exchanged "Enlightenment" ideas and discoveries. For all his scientific theorizing and philosophizing, he also was very much a realist who knew and understood dynamics among powerful people. Consequently, he showed little hesitancy in establishing connections with royalty, high government officials, and wealthy individuals, actively forming a useful network of international contacts, as he worked to create a career for himself and achieve his ambitions.

In the early 1770s, for instance, Volta began to correspond with Joseph Priestley, the Englishman who was a pioneer in modern chemistry, and more significantly for Volta, author of *History and Present State of Electricity*. Priestley classified his book as a compendium of current research on electricity, and Volta became familiar with this book when a French translation was published several years after the original English language version came out in 1767.

By writing to Priestley, Volta had more in mind than just wanting to share and exchange ideas of scientific content. Priestley initially promised updated versions of *History* based on the latest research and discoveries in the field, and Volta hoped to be referenced in additional editions. While no such updates ever materialized, Volta's letter-writing exchanges with Priestley helped him in other ways. One case in point occurred in 1773. Following his discovery of alkaline air, Priestley mentioned it to a select group of private individuals before making it public in March 1774. The English scientist wrote about this new finding to Volta in a letter dated November 10, 1773, just weeks after he'd brought it to Benjamin Franklin's attention.

Priestley came to hold Volta in high regard and helped his younger colleague develop connections in the important English scientific community that would prove significant when Volta traveled to England in 1782.

## HIS FIRST SUCCESS

Volta was just twenty-four in 1769 when he published his first treatise on electricity: *"De Vi Attractiva Ignis Electrici, Ac Phaenomenis Inde Pendentibus,"* or "On the Attractive Force of Electric Fire and Related Phenomena." In this seventy-two-page paper, written as a letter to Beccaria, Volta presented his ideas on electrical attraction. Although he made references to his experiments with generating an electrical charge by rubbing different metals together, these mentions were more speculative in nature.

Four years later, Volta published his second paper, *"Novus ac Simplicissimus Electricorum Tentaminum Apparatus,*" or "New and Very Simple Apparatus for Electrical Tests," which showed that he favored more practical investigations into electricity over philosophical ones. It would be an overstatement to say that these works launched him into the elite of the European scientific community. But these papers did raise Volta's profile—his description of his all-wood electrostatic machine, in particular, drew the attention of colleagues.

Professionally, Volta's next steps occurred primarily in the academic world, a field where natural philosophers could find financial support during that era. In 1774, Volta took an appointment as superintendent of Como's secondary schools, and the following year he started teaching experimental physics in Como's public grammar school. He held this job until 1778, at which point he became a professor of physics at the University of Pavia, where he taught for forty years.

His career as a professor was highly successful. His students were said to have idolized him, and he seemingly loved to teach. Volta remained a professor even after he found great professional triumphs, and after he stopped his active career as a scientific inventor.

Volta's first big career breakthrough arrived in 1775, when he announced the invention of the electrophorus, which means "electricity bearer" in ancient Greek. An early pioneering type of an electrical induction device, it could generate, as well as store, electricity. The machine retained an electric charge through the then-unique concept of "electrostatic influence" instead of the standard way during that era of direct electrostatic friction.

Historian Joyce Chaplin described how Volta's invention, sometimes called the electrophone, worked in this way:

*Volta's 'electrophore' sandwiched an electrostatic cake (a blend of turpentine, resin, or wax) between a fixed metal plate below and a rotating wooden shield covered with tinfoil above. When the cake was rubbed (while the lower plate was grounded), it generated and condensed a charge (like a Leyden jar), but its fixed metal plate also retained a charge—it did not decay.*

Volta wrote to Priestley to tell him about the electrophorus. He not only wanted to let his English colleague know about his latest invention, he also sought to find out from his knowledgeable mentor whether anyone else might have been working on a similar invention. In fact, the Swedish electrician Johannes Wilcke had actually invented a similar electrical apparatus in the 1760s. Volta, working independently of Wilcke, was totally unfamiliar with this other device when he created an electrophorus.

Surprisingly, Volta was not upset over this news. Instead he was quite pleased to find out about Wilcke's device because it provided additional support for his own concept of "spontaneous electricity." Furthermore, Volta's invention was viewed as improving upon Wilcke's because it brought a new, important ingredient to perpetual electricity devices. Volta's electrophorus was simple to construct as well as easy to transport, so it was welcomed by scientists looking to display their electricity experiments. Up until that point, electrical devices had been of the experimental variety, while Volta's had practical uses. As a result, he felt justifiably triumphant that he had devised a better version of this type of machine.

Volta was not shy about demonstrating his electrophorus device. One reason for this was because it took nearly a year for Priestley's reply to reach Volta. His demonstrations, consequently, served to publicize both his invention and himself. Volta showed off the electrophorus to audiences of important public figures, professors, and amateur scientists around the Milan area, and news of the invention eventually made its way to Vienna. The device was so well received that many were built through the rest of the year, including some as large as seven feet in diameter.

This invention gave Volta, who was then just around thirty years old, a major boost in stature within the scientific community. He had demonstrated intelligence and cleverness by inventing a device that improved upon previous advances like the Leyden jar and existing knowledge of static electricity. In doing so, he also challenged the prevailing theories of Benjamin Franklin, then a legendary figure in the science of electricity for inventing the lightning rod in 1752 and his subsequent contributions to this field of study. The electrophorus constituted a rejection of Franklin's notion that electricity was something fluid that could move within an equilibrium. Although the device may not have any practical purposes today, it played a significant role in eighteenth-century electrical research and in the evolution of our understanding of electricity.

# Chapter Two

## VOLTA IN CONTEXT

### THE TIME OF ENLIGHTENMENT

Alessandro Volta lived during the full flowering of the Age of Enlightenment, which typically is defined as having occurred from the late seventeenth century into the early nineteenth century. While historians don't agree on the precise dates for the Age of Enlightenment, they do agree on the tremendous importance that it had on the world—both during its time and in the centuries to come. In his introduction to the book *1001 Inventions That Changed The World*, noted science author Jack Challoner asserts that "the seventeenth and eighteenth centuries were a period of rapid progress in astronomy, biology, physics, chemistry, and mathematics . . . During this time, scientists also began to experiment with the two vitally important forces of electricity and magnetism."

The publications of Isaac Newton's *Principia Mathematica* (1686) and John Locke's "Essay Concerning Human

Understanding" (1689) frequently are used as examples for pioneering works of the Enlightenment. Newton's theories on calculus and optics certainly proved to be hugely influential on Enlightenment scientists' methodology involving precise measurements and illumination. Similarly, the interest in hands-on experiments also can be tied to Locke's idea that knowledge can be acquired through accumulated experience.

Volta's life and career very much embodied the Enlightenment era. Thinkers of this time placed great value on performing precise experimentation as well as setting high standards for empirical observations. Volta strongly believed in the importance of both principles. He relied on empirical evidence to prove his theories and engaged in dialogues with an array of scientists and philosophers to present his ideas.

Enlightenment thinkers agreed on similar tenets regarding rational questioning and the importance of dialogue to affect progress. During this period, "Science, previously erected on stilts of axioms and premises, was stripped to the ground," asserts noted American historian and political scientist James MacGregor Burns. "Change was at the very root of this new era . . . enlightenment and liberation raised men and women into a condition of possibility, the opportunity to better themselves and their world." The Age of Enlightenment created a fertile ground for innovation and invention by young scientists like Alessandro Volta.

## THE DAWN OF ELECTRICITY

The story behind electricity's discovery covers hundreds of years, stretching far back before the Age of Enlightenment. It features a gallery of curious characters, including a queen's doctor, a German mayor, and a French abbot.

The queen's doctor was Dr. William Gilbert, who was the physician to the first Queen Elizabeth from 1601–1603. Besides tending to his royal patients and serving as the president of the Royal Court of Physicians, Dr. Gilbert also held a fascination for magnets. While compasses had been used for several centuries by then, their magnetic powers had always been explained as a natural occurrence, with little scientific research done to explore how compasses and magnets actually worked.

After doing intensive research on magnets, Dr. Gilbert published a book, aptly entitled *On the Magnet*, in 1600, in which he outlined the theory of "electrical attraction" in discussing how magnetism worked. It was Gilbert who coined the word "electricus" to described substances that had these powers of attraction. One of these substances was amber, and "electricus" means "of amber" in New Latin (and "electron" means amber in ancient Greek). In *On the Magnet*, Dr. Gilbert included a drawing of a "versorium," a pedestaled piece of metal that moved when placed near a rubbed electric substance—the first illustration of an electrical device.

Dr. Gilbert's other major contribution to the field of scientific study was his reliance on empirical evidence, which he described in his book. Dr. Gilbert did not turn to hearsay and traditional ideas to draw his conclusions, but he derived his theories and explanations from discoveries that came about through his own experimentation. While this methodology of scientific research might be obvious today, it was unique for his time period, and his approach influenced scientists and inventors for generations to come.

Many electrical innovations that followed Dr. Gilbert's were done not by dedicated scientists but by men who dabbled in science. In fact, the term "scientist" was only first used in print in the 1830s. Electricians and natural philosophers are terms that were typically used to describe early experimenters in electricity (all three words will be used in this book).

Otto von Guericke was the burgomaster, or mayor, of the German city Magdeburg for approximately thirty years, and he was also an inventor. He devised the first vacuum-air pump in the 1640s, and the first water barometer around 1660. Later in the 1660s, he constructed the first electrical machine, although it only received widespread attention after being described and illustrated in a book that von Guericke published in 1672. Von Guericke's machine featured a sulfur ball that when rubbed and rotated would attract and repel everyday items such as paper, feathers, and other lightweight items. This "electrical generator" could create man-made electrical sparks, marking the first time this was done by a machine. While this device represents a significant advancement in electricity, it was mostly seen as a novelty, which was a common reaction during the early days of electrical study.

Francis Hauksbee's occupation was instrument maker. In the eighteenth century he made equipment such as telescopes and hydrostatic balances. This Englishman also was a talented inventor. He invented the double-barreled air pump, which marked a milestone advance over the air pumps that were based on von Guericke's style.

In the early 1700s, Hauksbee became intrigued with electrical friction after seeing the effects of mercury being shaken inside a glass tube. He built a hand-cranked friction apparatus involving ropes, pulleys, and a glass globe that was lined with wax, sulfur, or pitch. When he put his hand inside the globe, a luminous glow was created inside the glass. Hauksbee also could produce an electrical spark by removing his hand while the glass globe turned. He demonstrated this generator-type device before the Royal Society of London, but somewhat surprisingly neither Hauksbee nor his contemporaries were interested in harnessing these electrical charges for useful purpose.

During the first half of the eighteenth century, a number of electrical devices were created and mostly used for parlor tricks. One particularly theatrical demonstration involved placing an apple inside the glass receiver of the air-pump. By removing the air, the demonstrators created a vacuum, making the apple explode. The dramatic presentations by French scientist Pierre Polinière that demonstrated this effect were so popular that he performed them before Louis XV in 1722.

Another popular type of presentation was one that dramatized how far an electrical charge could travel. These demonstrations became quite elaborate after the invention of what is now known as the Leyden jar—a bottle that could temporarily hold electrons.

The Leyden jar was made in 1745 by a man named E. G. von Kleist, a German jurist and a Lutheran cleric who also liked to experiment with electricity. The device is called the Leyden jar because it became associated with Pieter van Musschenbroek, a University of Leyden physicist who made the same discovery as von Kleist the following year. The news of van Musschenbroek's findings spread more rapidly and widely than that of von Kleist's discovery, so it was named after the university in which van Musschenbroek taught.

The Leyden jar, typically described today as a capacitor, inspired a variety of entertaining scientific exhibitions. Igniting gunpowder, lighting candles, and killing small birds were all part of Leyden jar-based experiments.

William Watson, the English scientist credited with coining the term "circuit," was also known for his grand experiments. In 1747, for example, he ran 1,200 feet of wire across Westminster Bridge over the Thames River. The wire not only shocked the men holding it at each end on opposite sides of the bridge, but it also made alcohol shoot out of a bottle. The following year, Watson rigged up a large-scale demonstration that involved a cranked wheel that transported an electrical charge when a boy, who was suspended in the air, touched a girl while she was standing on a tar-covered tub.

Jean-Antoine Nollet, one of Volta's early mentors, was the Abbot of Paris's Grand Convent of the Carthusians, but as a scientific experimenter, he staged grandiose electrical displays. In 1746, he received an audience with King Louis XV at Versailles where he demonstrated to the king how an electrical discharge could run through a group of 180 Royal Guardsmen, while they were holding hands—the charge also made the men all jump from the shock. The amused king later had Nollet perform this electrical shock stunt again in Paris. In this second presentation, however, Nollet used 700 Carthusian monks, who all jumped when electricity jolted them.

Nollet's theatrics make Franklin's kite and lightning experiment seem rather mild in comparison. Franklin, remember, did not see eye-to-eye with Nollet. He was not a fan of the two-fluid electricity theory—the idea that a positively charged fluid and a negatively charged fluid could move through bodies to connect and create electricity—to which Nollet subscribed. Instead,

Franklin believed in the single electrical fluid theory, which asserted electricity acted externally, not internally, on bodies.

It should be noted that Volta was not above displaying a little showmanship in his own demonstrations. Indeed, he had a knack for presenting his experiments in dramatic fashion, which was beneficial when he demonstrated his battery to Napoleon Bonaparte in 1801. Volta had a reason behind engaging in theatrics when showing off his latest inventions to visitors in his lab or before a gathered audience—he wanted common people to become interested in the world of science. He thought that by making science fun and something that men and women could relate to in their lives, he could help popularize science. The eighteenth century did see an increased interest in science and Volta, in his own way, played a role in this development.

## THE INFLUENCE OF ENLIGHTENMENT

When Volta turned his back on a career in the church for one in science, it was not only a personal choice. It reflected a larger societal trend that blossomed during his lifetime. In the eighteenth century, universities shifted away from a church-influenced curriculum to one that was more secularly based. This shift often arose because governments sought to weaken the church's role in education. Not only was this exemplified in Volta's academic preference for science and the humanities over working for the church, but it was also reflected in his long career in academics, both as an educator (he served as a director of a school in hometown of Como) and as a professor (his more than forty-year teaching career, primarily at the University of Pavia).

The seventeenth and eighteenth centuries saw a significant

increase in the number of science societies and academies that focused on creating and disseminating scientific knowledge, more so than the mainly scholastically-oriented universities. The growth was so pronounced that this era has been dubbed "The Ages of the Academies."

Volta made much use of the science societies throughout Europe. These were places where he could connect with his fellow scientists and share his theories and inventions. In his comprehensive biography, *Volta: Science and Culture in the Age of Enlightenment*, Giuliano Pancaldi presents data demonstrating how Volta corresponded with more people in Europe than all of the other areas of Italy (outside of his native Lombardy).

Pancaldi compiled and analyzed statistics related to Italian scientists of Volta's epoch, and found that the majority of the seventy-four scientists he researched came from the professional class or, as in the case of Volta, lesser nobility. He researched the book collections in the libraries of several of Volta's peers—one scientist, one physics professor, and one government minister, who also was a Volta supporter—and found that all these libraries contained books that were published in European nations. This reinforces the idea that information really was able to spread beyond borders during the eighteenth century.

The push for intellectual dissemination was driven by the universities' wish to promote and enhance their reputations by sending their professors abroad. When professors for one area in Europe spent time in other places, their ideas and philosophies naturally spread. Volta enthusiastically participated in these trips since they presented an opportunity (and a sponsored one at that) to meet other scientists and promote his own work.

## THE EGALITARIAN REVOLUTION

Among the Enlightenment's core principles were that knowledge could transcend political boundaries and rival state powers, and that the public should be able to freely ask questions regarding governments and religious institutions. These egalitarian-rooted concepts were key tenets for an intellectual network of men called the Republic of Letters, and they played out in several aspects of society in the seventeenth and eighteenth centuries —and had both direct and indirect consequences on Volta's life and career.

The eighteenth century saw a rise in literacy, a weakening of censorship (Sweden and Denmark outlawed it in 1766 and 1770, respectively) and a shift in population from the country to urban areas. While perhaps not initially apparent, these three movements have ties to one another trend. The more literate citizens became, the more knowledge they acquired and the more aware they were of ways in which they were being oppressed, such as through censorship. Similarly, the more they read, the more they learned about events in the world, and cities offered people (particularly young people) opportunities, from employment to entertainment, that they could not find in the countryside.

These historical developments all served to aid in the Enlightenment belief in a broad transmission of ideas, a belief which spread greatly during that century, and continues to expand at ever-increasing speed today. The flow of ideas also served to develop interest in such fundamental Enlightenment topics as science, politics, natural philosophy, and the humanities.

These societal changes, moreover, played key roles in creating cracks in the traditional European class and political systems.

The French Revolution of 1789 is just one famous example of the consequences that arose from these societal fissures. This revolution not only struck a blow against censorship, it also epitomized the rise of the common man—creating a world where an ordinary genius like Volta, as a member of the lesser nobility, could circulate in elite scientific circles and upper class society, and achieve great career success.

The increase in literacy during the 1700s had a direct correlation to the tremendous growth in the book industry during this time period. Each fed the other. More than 640,000 books were published in Europe between 1751–1800, after fewer than 550,000 had been published in the entire sixteenth century. James Lackington, the Englishman credited with revolutionizing the book business with his massive London bookstore named The Temple of Muses, observed that "the poorer sort of farmers, and even the poor country people in general, who before . . . spent their winter evenings [with] stories of witches, ghosts, hobgoblins . . . now shorten the winter nights by hearing their sons and daughters read tales, [and] romances."

One sign of how much the book trade evolved during the eighteenth century can be seen in the jobs of the book publisher and bookseller. In the seventeenth century, these were basically the same job. The book business was still in its infancy; the man who made and published the book was also the one who sold the book. During the next century, tremendous growth in the book business resulted in the book publisher and the bookseller becoming two separate enterprises.

The 1700s saw the dawn of illustrated magazines and lithographs, along with printing press innovations such as the first color engravings. Additionally, there was a large increase in books printed in a variety of languages; previously most books were printed in Latin. Many historians have credited this language shift with the rise of eighteenth-century nationalism.

Books, without question, also reached a far wider spectrum of social classes during the 1700s. Not only did this growth in book publishing satisfy the increasingly literate society's interest in ideas, but exemplified the greater importance on learning during this era, coinciding with the importance that Enlightenment thinkers placed on education.

Knowledge, according to John Locke, comes from our experiences. With the increased emphasis on education and the related rise in literacy during the eighteenth century, it should come as no surprise that more people wanted to travel and discover the world around them during this time period. The 1700s, in fact, became the golden era for the Grand Tour, where citizens embarked on long trips to visit a range of destinations primarily to learn about those foreign places.

This type of travel served to advance the cultural exchanges of ideas and information between men and women who lived in lands far apart from each other, whereas previously they had few opportunities to interact. Letter writing was the predominant form of communication and was limited by language barriers and individual literacy.

Traveling became easier in the eighteenth century because train lines became more commonplace. The 1700s saw improvements both in train engines and in rails, which contributed to the rise in railway travel. Some historians also consider the increased popularity in railroads a result of a sad consequence of the huge number of horses that were killed during the seemingly endless wars fought in the eighteenth century; the French Revolution and the Napoleonic Wars, in particular, resulted in high death tolls for horses.

The transnational developments in Europe during the eighteenth century also reflected the Republic of Letters' philosophies, which held a commitment to scholarship and intellectual discussions unrelated to local politics and authorities. Volta naturally was drawn to the ideas of the men who were part of the Republic of Letters. Their principles matched his own, from his interest in disseminating ideas beyond national borders to his desire to be part of a cosmopolitan network of experts. By associating with this community of scholars, Volta joined with other natural philosophers and scientists who sought out their peers for inspiration, judging, and reward for their isolated work.

## PRAGMATISM AND THE UNCONVENTIONAL THINKER

Volta exhibited sympathies for secular Enlightenment theories and values since he preferred scientific rather than religious explanations of natural phenomena. He made sure, however, to keep his more anti-establishment views regarding society, religion, and philosophy private. He portrayed himself as an innocuous, if somewhat unconventional electrician, and avoided getting mixed up in contentious debates about politics and religion.

No place was Volta's conventional approach more apparent than in his attitudes toward religion—namely Roman Catholicism, the predominant faith of Italy then and now. Perhaps because so many members of his family made their living through the Church, Volta always displayed in public a respect for Catholicism, the Church, and its practices.

Volta also had to avoid attracting controversy when he traveled as a representative of his university or his patrons.

He could not risk upsetting the people who bankrolled his trips to the point where they might withdraw their support. As a professor, moreover, Volta derived his primary income from a state appointment. While his fame and success earned him a certain degree of security in his later years, he did not have significant financial autonomy in his life. This situation was especially true at the beginning of his career, when he needed to toe the line to receive support—financial and otherwise—so that he would be able to continue his travels.

# Chapter Three

## VOLTA THE TRAVELER

### THE POLITICAL CLIMATE OF
### EIGHTEENTH-CENTURY EUROPE

Europeans, during the eighteenth century, endured political conflicts, wars, and revolutions. There was the Seven Years War and the one-day Kettle War. Four different Wars of Succession were fought: the War of the Spanish Succession (1701–1713), the War of the Polish Succession (1733–1738), the War of the Austrian Succession (1740–1748), and the War of the Bavarian Succession (1778–1779). The Turks fought battles with Austria and Russia three times during this century. Perhaps the most notable was the French Revolution, which changed the course of world history, but also played a significant role in Volta's own world.

Four major political powerhouses dominated Europe during the 1700s. The Hapsburgs ruled Austria and controlled the Holy Roman Empire for most of the century. The formidable Queen

Maria Theresa, who held the Hapsburg throne from 1740–1780, ruled with a style of enlightened absolutism that was borne from being practical-minded.

The House of Bourbon held reign over France, with the King Louis trio (Louis XIV, Louis XV, and Louis XVI) demonstrating a determined resolve to strengthen France's place in a world. Prussia's power was in the hands of the Hohenzollern family. Their most prominent king during this era was Frederick II. Known as Frederick the Great, his forty-six years as king (1740–1786) was the longest of any Hohenzollerns. While quite aggressive in expanding Prussian territory, Frederick was also a patron of the arts and he considered himself to be something of an enlightened ruler.

In the early 1700s, control of Great Britain shifted from the House of Stuarts, who had held the throne for several centuries, to the German House of Hanover. After taking over the crown, however, the Hanovers had to contend with much internal unrest as well as having to deal with ongoing conflicts with France. It was King George III, the third of four King Georges in England during the eighteenth century, who was on the throne during the American Revolution and the French Revolution.

Italy, meanwhile, was not a nation during the 1700s, but more of a collection of states. The process of unifying these states into the modern nation of Italy did not start until 1815, and would not be completed until the second half of the 1800s. During the eighteenth century, control of the Italian states was often traded by the major European powers. Lombardy, however, was fortunate to exist with a relatively stable political environment during this time period. Austria ruled the region for most of the century until Napoleon-led France gained control in 1797 (although Austria returned to power in 1815).

Besides engaging in many military battles, European nations also vied with each other, to use Pancaldi's term, in "the imitation-competition game" during the 1700s. Intellectual competition resulted in governments backing initiatives to support programs involving culture and education. By improving their academic institutions, for example, rulers could enhance their own reputations across the continent as well as within their own country.

Science was one area that benefited from this agenda. Governments wanted to enrich the quality of their scientific collections along with showing off their nation's scholarly talent. They would frequently utilize academics and other experts to represent them abroad. This type of diplomacy today is often described as "soft power."

In 1777, the Austrian Empire's public administration developed a plan for state-supported trips that focused on enhancing science instruction and technology. The program connected the empire's top professors with their fellow intellectuals in other countries for the purpose of boosting Austria's reputation. Austrian public administrators not only wanted their notable academics to serve in a promotional function for the nation, but to act as scholar-agents who could observe advances being made abroad and bring back useful information to improve Austria's standing in this "imitation-competition game."

Not surprisingly, Volta was quite eager to participate in this program. Because he was essentially self-trained and lived in an isolated small city, Volta viewed travel as an exciting opportunity to advance his career. By visiting other countries, he could meet other natural philosophers and potential patrons, while also learning about the latest inventions. He'd corresponded with his fellow scientists, but this was not the same as actually meeting, and hopefully impressing, them face-to-face. Furthermore, Volta's professional status could rise by being part of Austria's Lombardy public administration, and that would create better possibilities of taking his career to a higher level.

These state-organized promotional trips made sense monetarily too, because governments allocated discretionary funds to support these "soft power" expeditions that otherwise would have been difficult for Volta to finance on his own. These trips wound up being the very definition of a "win-win" for Austria and Volta.

During the spring of 1777, Volta began making his case to the Austrian government for a stipend for state-supported travel. Having achieved some renown abroad by this point for inventing the electrophorus, Volta successfully attracted the interest of the Austrian administration, which wanted to leverage his status to their advantage by sending him to other countries. By combining this pseudo-government agent role with being an up-and-coming scientist, Volta developed into "a skillful correspondent, a consummate traveler, and a good scientist-diplomat," as biographer Pancaldi puts it.

Representing the Austrian Empire gave Volta the means and opportunities to pursue his own professional activities in Europe. He could actively cultivate an array of business relationships by meeting with his fellow natural philosophers as well as courting influential people outside of the science world.

To some extent, Volta could even decide his itineraries for these trips. It is not surprising then that he picked regions in Britain, France, Holland, Switzerland, and the German states—

places that would be greatly beneficial for him to visit since they had strong scientific communities. Although he was representing Austria, Volta was an Italian and could present himself as a somewhat neutral party in a continent overflowing with political conflicts.

Additionally, these official trips afforded Volta opportunities to attend lectures, socialize with colleagues, visit laboratories, and observe the recent discoveries and scientific advances. As part of his duties on these travels, he met with instrument makers to learn about the latest lab equipment, and even possibly order items.

He could also play tourist. Exploring Europe's fabled cities, visiting museums, observing architecture, and enjoying the natural beauty of these foreign lands broadened his knowledge of the world and stimulated his keen intellect. Volta could find inspiration for his work in the most surprising places.

## VOLTA'S FIRST TRIP ABROAD

To secure his first sponsored trip, Volta had to vigorously lobby the powerful Austrian minister, Wenzel Anton Kaunitz. When he eventually approved of this request, the Vienna-based minister instructed the Austrian representative in Milan to remit the requisite funds to Volta for him to take a "literary trip." That Kaunitz authorized Volta to participate in this significant government program exemplified Volta's growing stature and reputation.

Kaunitz, however, budgeted only fifty zecchini—a gold coin that was worth approximately fourteen lira—for this excursion. This left Volta with limited resources of time and money for his trip, so he had to plan it in an economical way. He narrowed the choices for his destination to Austria and Switzerland.

Austria had the advantage of being the location of the empire's administrative headquarters, so Volta could meet with important people there. Switzerland, on the other hand, offered the opportunity to visit noted scientific institutions and meet with prominent natural philosophers who lived some distance from Lombardy. After consulting with one of his advisors, the Vienna Court Dignitary Baron Sperges, Volta decided on Switzerland.

In early September of 1777, Volta embarked his first trip abroad. Among his traveling companions were his longtime friend, Count Giambattista Giovio, and the philosopher and educationalist Francesco Venini. It is easy to imagine Volta's excitement as they set out trekking over the Alps. Ever since childhood Volta had seen the Alps towering over his hometown of Como. The mountain range stood between him and the rest of Europe, and now he had the long-awaited chance to cross the range and see the world outside of Lombardy.

When they visited the lake towns of Geneva and Lugano, Volta stopped to make some meteorological observations. In Zurich, the noted Swiss poet and painter Salomon Gessner hosted Volta, who had the chance to examine the extensive natural history collection assembled by Salomon's ancestor, the pioneering Swiss scientist Conrad Gessner in the sixteenth century.

This Swiss city was where Volta experienced a personal highlight. On September 16, 1777, he wowed a Zurich audience with a variety of experiments, which included an exciting—and explosive—demonstration of his methane gas pistol, a type of air gun used to spark a eudiometer. The presentation helped to bring

Volta some attention within the science community. The well-regarded Swiss lawyer, Johann Rodolph Valltravers, reported on Volta's Zurich exploits to none other than Benjamin Franklin. When he visited Basel and the Alsatian city of Strasbourg later on this trip, Volta demonstrated his hydrogen gas-burning lamp that ignited with an electric spark.

Much of Volta's time was spent meeting with scientists, naturalists, librarians, instrument makers, chemists, and physicians. He performed his duties as an emissary, and then made professional connections that would prove beneficial.

While in Basel, Volta met with the esteemed Swiss mathematician and physicist Daniel Bernouli. During his stay in Strasbourg, he was introduced to the politician Barbier de Tinan, a science enthusiast who liked to use his own electrical machine to make his guests' hair stand up through static electricity, and Baron Frederic Dietrich, who later played a major role in Volta's career when he presented several of Volta's experiments involving inflammable air to the Paris Academie des Sciences.

In Bern, Volta had the chance to visit with venerated scientist Albrecht von Haller, who would pass away before the year was out, along with meeting naturalist and philosopher Charles Bonnet, botanist Jean Senebier, and Horace Bénédict de Saussure. H. B. de Saussure was the geologist and electrical experimenter who, a few years later, invented the hair hygrometer, a meteorological device that for years was used to measure humidity. Making acquaintances with Senebier and de Saussure proved to be particularly important—the two men eventually became Volta's champions and advisors.

Following their stay in Switzerland, Volta and his companions traveled over the border into France and ventured to Ferney. There they were thrilled to have a half-hour audience with the eminent writer, and Republic of Letters icon, Voltaire—the nom de plume for writer and historian François-Marie Arouet.

The legendary satirist, it is said, made a grand arrival in an opulent carriage led by two mounted heralds. Volta matched wits with the eighty-year-old writer, who spoke to the young scientist in Italian.

Volta greatly enjoyed his first journey abroad, finding it both productive and stimulating. In fact, he would ultimately visit Switzerland five times. His friend Giambattista Giovio offered this observation of Volta during this trip:

*My Volta is always busy. What an industrious scholar he is! When he is not paying visits to museums or learned men, he devotes himself to experiments. He touches, investigates, reflects, takes notes on everything.*

The trip delivered short- and long-term benefits for Volta. The favorable impressions he made enhanced his already growing reputation. Men he met with, such as Deitrich, disseminated Volta's views and experiments. Scholars, such as Barbier de Tinan, would later translate Volta's works into several languages.

His trip also drew praise from his patron at the time, Carlo di Firmian, Austria's plenipotentiary minister in Milan. In a letter sent to Volta, Firmian stated that the contacts the young science professor made during his trip added to the "communication of enlightenment" and his information on pertinent discoveries were "unfailingly favored." Additionally, Firmian agreed with Volta's suggestion that funds should be allocated to purchase machines in Paris and Geneva for the physics cabinet at the University of Como.

## VOLTA'S 1781–1782 TRIP TO ENGLAND AND FRANCE

For his next trip, Volta sought to take a more extensive journey that related to his plan to submit an essay to the Royal Society of London. This goal of having the Royal Society publish his work followed the example set by scientists he admired, such as Franklin, Beccaria, and Priestley. To obtain government backing for this travel plan, he contacted Prince Charles of Lorraine, the most notable aristocratic patron he had at that time. Volta also used his position as professor at the University of Pavia, to which he was appointed to in 1778, as leverage to make this trip possible.

Volta's proposal did not get an immediate approval, so he agreed to travel to Tuscany in September 1780. This excursion, which was both Firmian-endorsed and government-supported, was funded with approximately double the budget that Volta received for his earlier journey to Switzerland. The "imitation-competition game" was the purpose behind this short trip to central Italy. Volta and Abbot Re, his instrument maker at the University of Pavia, were given the mission of studying a number of machines in both public and private cabinets located in Tuscany.

Volta had an ulterior motive in accepting this trip—it presented the opportunity for him to meet with Prince Cowper, a Fellow in the Royal Society in Florence. In fact, their meeting in Tuscany turned out to be quite productive, with the Prince ordering a eudiometer that was built according to Volta's designs. Volta also made several successful recommendations to Firmian regarding machines for the University of Pavia. One set of machines ended up being ordered through Jean Hyacinthe de Magellan, a major figure in the European scientific instrument market. De Magellan had become Volta's prime contact in England after Priestley had put the two in contact with one another in 1776.

After the success of his Tuscan business trip, Volta finally received permission to travel to England and France in mid-1781 when he was thirty-six years old. The trip was expanded to last more than a year in duration, and represented an important personal opportunity for Volta. He wound up meeting many professional peers whom he had known only through correspondence, and was able to expand his connections with aristocrats, scientists and others in Western Europe. It also proved to be a learning experience that helped him to grow as a physicist.

Volta's travels did not always go easily, as he was hindered by many complications, the most significant being lack of finances. The Austrian government agreed to provide only one-sixth of the funds needed for this trip. To make up the rest of the travel budget, Volta turned to his family, and came to a monetary agreement with his brother Luigi. He also defrayed some of the travel expense by spending a good deal of the trip accompanying the wealthy Marchioness Leonora Villani, her son, and a colonel.

Setting off in early September 1781, the group's first visits were in Switzerland, Germany, and Alsace. At these stops, Volta conducted official business of inspecting the science communities, but he experienced some early highlights. A visit to Belgium, for example, resulted in an invitation to the royal court of Brussels, and the chance to converse for an hour with the nation's monarchs.

Another early high point for Volta came during his time in Holland. In Haarlem, he met Martinus van Marum, the director of the Teylors Science Museum, home to the largest electrical machine then in existence. Van Marum subsequently shared his carriage with Volta when they traveled to Amsterdam, and thereafter the two maintained a correspondence. It stands as no coincidence that Volta received a quick acceptance into the Dutch Society of the Sciences.

Volta also spent a few days in Rotterdam with Prince Gallitzin, the Russian ambassador to Holland who was also an electrical experimenter. In fact, the Prince designed one of the biggest electrostatic machines of his day.

Volta's visit to Rotterdam left a very favorable impression. He described the city in this way:

*Rotterdam is a large and fine town, crossed by many navigable canals . . . All the buildings are extremely clean: everything shines like crystal; and indeed most of what is offered to view is polished crystal, because windows occupy a much larger share than walls on the exterior of houses, magazines, and shops. Everything is marked by an appearance of comfort that instills joy. Open roads are large, straight, splendidly paved, and large trees have been planted on both sides, to the right and left of every canal. Bridges, some of which are very fine, are many and the sight of people flooding everywhere—among whom you do not see a single man in rags—provides the best possible form of recreation in the world.*

Volta's description of Rotterdam reflects his thoughts overall about Holland. He ranked that nation along with England as countries that came closest to reaching his ideal of an enlightened, civilized nation. The qualities Volta said that he valued the most in a nation were "a well-run public administration, lively commerce and cultural pursuits, substantial secularization of the state and society, [and] a reasonable balance between individual riches and the welfare of the lower classes."

France was Volta's next destination. From January to April 1782, he stayed in Paris, making attempts to break into the ranks of France's scientific elite. He frequented lectures that featured such leading scientists as chemist Balthazar Georges Sage and physicist Jacques Alexandre César Charles. Volta also became acquainted with several important figures in the French science community, men such as the naturalist Count Buffon, the astronomer Pierre-Simon Laplace, and the chemists Claude Louis Berthollet and Antoine Lavoisier. Among the most notable men Volta met was the famous Benjamin Franklin—he even took a lunch at Franklin's residence.

Besides circulating in scientific circles, Volta also attended salons as part of his duties as a visiting scientist-diplomat. At these private gatherings, he mingled with members of the nobility class and high-ranking public officials. Some of the attendees were amateur experimenters. One such person was a Monsieur Le Noir, the head of the Paris police force who became a Volta fan and even invited the visiting scientist to his house on several occasions. Furthermore, Volta presented a few lectures on chemistry and electricity to Le Noir's daughter, Madame de Nanteuil.

Volta's time in Paris unfortunately was not the rousing success that he had hoped for. Although he collaborated with Laplace and Lavoisier on some experiments involving electricity and evaporation, this work did not lead to significant recognition in the science world. Adding to this disappointment was the fact that the Parisian scientists declined to acknowledge Volta's paper on his experiments in electricity and evaporation, which he had advertised in several places, including *Philosophical Transactions*. Despite this snub, Volta's advancements with electrometer sensitivity and the measuring techniques that he developed and used in these experiments would play a key role for him in future work.

Volta did have the opportunity to address the Académie Royale des Sciences on several occasions to discuss his experiments on electrical atmospheres and the condensatore (a condenser that could measure electricity), but he did not earn consideration during his stay for candidacy into the Académie. Even though Lavoisier co-wrote a letter later that year nominating Volta as a corresponding member of the Académie, Volta still failed to receive formal recognition from his French peers. Volta, however, was admitted into the less prestigious Vieux Musée de Paris.

But he did receive recognition in England when he arrived in there at the end of April. Preceding his visit to the land of Isaac Newton, there had been presentations—at the urging of Lord Cowper—of Volta's memoir on conjugate conductors over four separate meetings of the Royal Society in London.

Working to Volta's great advantage was that he arrived accompanied by his friend Jean de Magellan, who had rendez-voused with Volta on several occasions earlier during this trip. Magellan would play a critical role for Volta during his stay in England. Besides being a member of the Royal Society, Magellan also belonged to numerous clubs for instrument makers, natural philosophers, chemists, and physicians. Consequently, he served as a conduit for Volta, introducing him to many important members of the British science community. This chance to easily fraternize with, and learn from, the Britain's best natural philosophers became even more significant since the Royal

Academy did not convene a meeting during the months Volta was in England.

Volta spent nine weeks in London, and then continued for another three weeks touring England with Magellan. During his visit, he had the chance finally to meet two of his long-time correspondents, Joseph Priestley and Tiberius Cavallo, as well as such eminent British scientists as James Watt and Sir Joseph Banks. At the Portsmouth naval base, Volta was the guest of Admiral Lord Howe aboard the warship *Regina*. He visited industrial factories of northern England, and came away particularly impressed with the textiles mills in Manchester and Birmingham's metal industry.

Volta left England with an extremely high regard for the country, as he shared in a letter:

*Far from finding in England the decadence and weakness some speculate on, one sees in it nerve and vigor to an extent that no other country displays. Commerce seems to be increased, and gold certainly circulates very quickly. The riches of individuals are huge; the class of comfortably-off people is extremely large; the worker is well dressed, better fed, and despite taxes (such as to scare anybody) still earns enough money to throw it away in the taverns. Building, manufacturing, and new enterprises flourish everywhere . . . Those who know it can tell from its appearance and internal motions that the body is healthy, sturdy, well fed, rich of juice and blood.*

His affection for England was reciprocated with recognition. With support from Priestley, Volta reworked and published in *Philosophical Transactions* his lengthy memoir about conductors. This publication further strengthened his ties to England. When

Volta later addressed the Royal Society on his contribution to Galvanism—the effect of electrical current on a muscle—it was quite well received, and he was elected to the prestigious Society in 1791. Three years later, the Royal Society conferred upon him the prestigious Copley Medal, based on his research on galvanism.

## TOWARD THE GERMAN STATES

While England won his affections, Volta was pushed back to Vienna and the German states. During his time in London, his main patron, Carlo di Firmian, had passed away. His death followed that of Empress Maria Theresa's two years earlier and the rise of her son, Joseph II, to the position of ruler of the Austrian Empire. Volta realized that he needed to turn his attention toward Vienna, especially because Prince Kaunitz, the prime minister under Joseph II, took a hands-on approach in his oversight of the University of Pavia.

As it turned out, Kaunitz's new supervision brought both benefits and stipulations for Volta. These consequences were realized in Volta's next trip—a 1784 journey to Austria, the Czech regions, and the German states that he took with his fellow University of Pavia professor, the anatomist Antonio Scarpa.

By representing the court of Vienna, Volta received generous funding from the government for this four-month trip; it was much more than he received for his prior year-long trip. At the same time, the Prime Minister did not give Volta carte blanche for this excursion. During a meeting in Vienna, Kaunitz outlined his expectations, asserting that he had to focus more on "German literature," along with increasing his contacts and interactions with scholars from the German-speaking world.

Volta readily agreed to Kaunitz's ground rules since he needed a new patron.

Volta spent four months traveling through Austria, Saxony, Bohemia, Prussia, and Germany. Of the many German-speaking cities he visited, Göttingen and Berlin particularly stood out. During his sixteen-day stay in Berlin, Volta spent time with the renowned scholar Giuseppe Luigi Lagrange. Volta was surprised and delighted to discover that the Italian-born mathematician knew a great deal about physics and chemistry, and the two performed a variety of scientific experiments over the course of several evenings.

In Göttingen, Volta met the German scientist Georg Christoph Lichtenberg. The two men appreciated each other's scientific acumen, and Volta was impressed with Lichtenberg's extensive library, especially a German science textbook that he saw. It was also in Göttingen where Volta ordered German books and scientific instruments for the University of Pavia. In fact, the university's physics program eventually incorporated a great deal of German-language textbooks and instruments from German-speaking countries into its curriculum following Volta's visit.

During Volta's visit to Vienna, he became friends with the prominent mineralogist Baron Ignaz Edler Born. Born, who had a variety of political and cultural interests, was a supporter of secularized philosophy, as was Joseph II. From the baron, Volta realized that natural philosophers could co-exist with governments in both the administration of educational institutions as well as the spread of ideas that endorsed the government's policies. Born apparently had quite a sense of humor; he published a satire of the Church that Volta enjoyed immensely and shared with friends when he returned to Italy.

The significance of Volta's journey through these Germanic

states was that it earned him greater standing and respect from his Austrian government superiors for his work representing the empire as a scientist-diplomat.

Volta served the Austrians until 1796, when General Bonaparte and his French soldiers ended Austria's ninety years of rule in Lombardy. Because of his experiences dealing with government officials, Volta knew how to adjust and to compromise when dealing with the change in his homeland to French rule. It helped that Napoleon and his administrators did not hold radically different philosophies from the Austrians in regard to the roles of academics and natural philosophers in their governments. Consequently, the switch from Austrian to French rule was a relatively seamless for Volta. Furthermore, he developed a unique patronage relationship with Napoleon.

# Chapter Four

## THE METHANE EFFECT

### GONE FISHING

Volta's long expeditions around Europe certainly provided many benefits to his career, but one of his shortest trips paid unexpectedly big dividends.

In 1776, Volta went on a boat outing to Lake Maggiore, located not far from Como along the Swiss-Italian border. Lake Maggiore is Italy's longest lake and offers a very lovely spot for a relaxing fishing trip. Volta's outing there, however, was not for recreation; he went there with work in mind.

Volta's friend, Father Carlo Giuseppe Campi, had told him about a visit to Lake Maggiore during which he noticed that bubbles surfaced in the marshy waters of a spring would seemingly catch fire. The two men planned to go to the lake and investigate this unusual occurrence together, but this trip got canceled because Volta became sick.

Father Campi's story, however, continued to intrigue Volta.

He wanted to see for himself this bubbly phenomenon that his friend had described. Stoked by his exceptional curiosity, Volta eventually planned a trip on his own to Lake Maggiore so that he could explore the spring, which was located near the town of Angera.

While rowing his boat on the lake, Volta glimpsed bubbles rising to the water's surface. He noted that they appeared mostly where the lake was shallower and marshier, with more bubbles arising when the water at the bottom of the swampy area near the shore was agitated.

Volta did not know what he had discovered, but he knew that he had found something significant, so he collected some of this gas from the lake's surface to take back with him. As an admirer of Benjamin Franklin, he had read Franklin's paper on marsh gas, which served to inspire him to investigate this similar occurrence of "flammable air."

## BENJAMIN FRANKLIN

Our current understanding of methane owes much to yet another trip—one that Benjamin Franklin took in 1764 through New Jersey. In a 1774 letter to Joseph Priestley, Franklin described what happened on this trip a decade earlier:

> *When I passed through New Jersey in 1764, I heard it several times mentioned, that, by applying a lighted candle near the surface of some of their rivers, a sudden flame would catch and spread on the water, continuing to burn for near half a minute. But the accounts I received were so imperfect, that I could form no guess at the cause of such an effect, and rather doubted the truth of it. I had no opportunity of seeing the experiment; but,*

*calling to see a friend who happened to be just returning home from making it himself, I learned from him the manner of it; which was to choose a shallow place, where the bottom could be reached by a walking-stick, and was muddy; the mud was first to be stirred with the stick, and, when a number of small bubbles began to arise from it, the candle was applied. The flame was so sudden and so strong, that it [caught] his ruffle and spoiled, as I saw.*

This experiment remained on Franklin's mind when he returned to England at the end of that year. He discussed it with some learned friends there, but they didn't take the idea seriously. "I suppose I was thought a little too credulous," Franklin once wrote about his colleagues' reaction. Still, Franklin cared enough about marsh gas to conduct additional experiments in England. These experiments, however, didn't turn out so well, as Franklin caught a fever from his mucking around in unclean water.

Franklin's research was taken more seriously back in New Jersey. No less a personage than Dr. Samuel Finley, the president of the College of New Jersey (now known as Princeton University), wrote to the Royal Society in 1765 about the experiment.

In his letter, Dr. Finley described the experiences of a local man who had set marsh gas on fire and that he, himself, repeated the experiment.

*A worthy gentleman, who lives at a few miles distance, informed me, that in a certain small cove of a mill-pond, near his house, he was surprised to see the surface of the water blaze like inflamed spirits. I soon after went to the place, and made the experiment with the same success. The bottom of the creek was muddy, and when stirred up, so as to cause a considerable curl on the surface, and a lighted candle held within two or three inches of it, the whole surface was in a blaze, as instantly as the vapour of warm inflammable spirits, and continued, when strongly agitated, for the space of several seconds. It was at first imagined to be peculiar to that place; but upon trial it was soon found, that such a bottom in other places exhibited the same phenomenon. The discovery was accidentally made by one belonging to the mill.*

While the letter was read at the Royal Society, it failed to qualify for publication in the Royal Society's *Transactions*. As Franklin reflected years later, "It was thought too strange to be true, and some ridicule might be apprehended, if any member should attempt to repeat it, in order to ascertain, or refute it."

Without the Royal Society's imprimatur for this marsh gas research, Franklin moved on to other topics. During a trip to London in 1774, however, he met with Joseph Priestley. Priestley was captivated by the marsh gas phenomenon, and asked Franklin to prepare an entry about it for his upcoming book, *Experiments and Observations on Different Kinds of Airs*. Accepting Priestley's invitation, Franklin wrote a letter in the spring of 1774 about marsh gas, now known as methane. A couple of years later, Volta did his investigation at Lake Maggiore.

Franklin, of course, played a huge role in the story of electricity. He was the person who coined the terms "positive" and "negative" for electrical charges, and devised the "one-fluid" theory of electricity (electricity involved in the movement of a single liquid). Then there was Franklin's famous experiment in 1752 with flying a kite in a thunderstorm, one of the best-known science stories in history. This experiment very much reflects the Enlightenment belief in hands-on experimentation rather than theoretical concepts. Franklin's scientific interest in weather was shared by Volta.

## THE GASEOUS WATERS OF LAKE MAGGIORE

"Since ancient times combustible gas has been known to seep from geological fissures in certain areas of the world," writes historian Ralph S. Wolfe. "However, the experiments of Alessandro Volta with combustible air obtained from sediments and marshy places created widespread interest and laid the scientific foundation for study of the biological formation of methane . . ."

After conducting experiments with the substances that he had collected in Lake Maggiore's marshes, Volta described his discoveries in a series of letters to Father Carlo Campi. This correspondence not only reveals what Volta learned in his experiments, but also provides an insightful look at his methodology.

Volta's first letter to Father Campi offers a wonderful first-person description of what he observed on Lake Maggiore:

*So, on the 3rd of this month, with my head full of such ideas and being in a little boat on Lake Maggiore, and passing close to an area covered with reeds, I started to poke and stir the bottom with my cane. So much air emerged that I decided to collect a quantity in a large glass container.*

Later in his letter, Volta writes a thorough account of his scientific approach:

*Now for some details. This air burns with a beautiful blue flame. To make it burn and to produce the flame, the mouth of the vessel must be wide. If it is too narrow, when one puts a little lighted candle close to it, one hears many slight explosions in rapid succession, so slight that they can barely be heard.*

*I usually use, for simple experiments, a little container made of glass, cylindrical in shape, three or four thumbs long, and the same width except at the mouth which is approximately ½ thumb wide. When a candle is put close to the mouth, it is pretty to see how it gets covered with a small blue flame, which descends slowly along the walls of the vessel, almost as if licking it, until it reaches the bottom. But the show is more beautiful and more curious if we put a bit of lighted candle inside the container, using a bent wire; then the blue flames enlarge and increase in vigor. If the candle is lowered too far, it goes out, while at the mouth the air continues to burn. Then if the candle is moved away from the bottom, it will light again as soon as it touches the flame at the rim.*

*Isn't this the same thing that happens to the alcohol of wine? A torch immersed in such a fluid goes out, but when it approaches the surface it ignites and burns brightly. What better proof can there be that this flammable air, the same as any combustible substance, cannot burn unless it comes into contact with the ordinary air of the atmosphere?*

These passages provide great examples of Volta's deductive reasoning and his innate knack for presentation. He describes the logic behind his reasoning, along with a discussion of the quality of "the show." While only thirty-one years old at that time, Volta already understood that much of the power of science came from its visual appeal—the "wow factor"—and he clearly knew how to leverage it to his advantage.

His writing makes it easy to visualize how he selected the glass tubes with care. Using the simple tools that were available to him at that time and place, Volta proceeded to work diligently toward his discoveries. Every time he brought a lighted candle near to the combustible air, he developed a more accurate understanding of this unfamiliar phenomenon. Each lighted candle and each blue flame brought him closer to a breakthrough.

Volta's already extensive experience studying nature, coupled with his keen sense of ambition and a spirit filled with the potential of the Enlightenment, made him well aware of the importance that these experiments held.

In his second letter to Father Campi, he continued his discussion of what happened on Lake Maggiore, and revealed his experimenting process:

*After having tested the soil that sleeps, as to say, under the water, it occurred to me to examine the soil near the water but not wet. For this, I chose a marshy soil left almost dry by the subsiding of our Lake, and got ready to make the test in two ways. The first was to dig a few little holes in the mud (others were already prepared by deep animal tracks) and when they had filled with water, I stirred the bottom as usual with my cane, and let the air escape. I carefully collected some of it, and it did not fail the test; it caught fire.*

*The other test which gave me a more beautiful and charming sight was to push my cane in to the depth of about a foot in a place where the soil was softer and blacker, and covered with rotted grass, then pull it out all of a sudden and at once push a small lighted candle right next to the opening. It was just beautiful to see a blue flame appear at once and part of it fling itself upward while part of it went deep into the hole and touched the bottom.*

*Then I hurriedly dug several more holes close together, and I could not get enough of watching the flame run from one to another, setting one and then another ablaze, then all of them burning and shining together. But if I pressed the ground with one foot, or trod upon it so that more air rushed out, only some of the holes would burn . . .*

This letter shows Volta's excitement in the process of discovery and exploration. His prose may be antiquated by today's standards, but his exuberance and exhilaration—the joy he derived from observing scientific wonders at work—certainly shines through.

In his second letter to Father Campi, Volta arrived at the conclusion that this air could be powerfully combustible:

*No, sir, no air is more combustible than the air from marshy soil. In the first place, we can deduce this from the extraordinary number of small explosions we can get from it. But a surer indication is that it transmits the property of flammability to the ordinary air with which it comes into contact, and in this respect it far surpasses other combustible air. The strongest of all these, obtained by dissolving iron filings in vitriolic (sulfuric) acid, makes the loudest explosions when combined with a volume of ordinary air twice its own. The air of swamps, on the other hand, ignites and explodes most loudly, if to one part of it, we add 8 or 10 parts of ordinary air. If to one part of it, we add only 5 or 6 parts of ordinary air, it does not explode with its maximum flash and roar, but keeps flashing with a succession of small flames; finally, if we increase the proportion to twelve to one, the swamp air sets afire the whole mass.*

*Now we can understand why this swamp air burns so lazily in containers and why it is imperative for their mouths to be wide. No, it is not lack of flammability; it is rather an indication of excess of flammability; since in order to burn more brightly, it must be diluted with ordinary air.*

These observations represented an epochal discovery. In the letters that he wrote to his friend Campi, Volta accurately explains that inflammable gas comes from decaying vegetable substances that existed in marsh settings. This revelation was recognized as significant, with several of these letters translated into German and French.

## VOLTA AND THE LEGACY OF METHANE

Through his experiments at Lake Maggiore, and at other marshes in Lombardy, Alessandro Volta discovered, isolated, and identified methane, then known as marsh gas, between 1776 and 1778. "Volta concluded correctly that sediment 'air' from marshy soil contains more energy than does hydrogen," Ralph S. Wolfe observed, adding that "indeed, the methane from marshy soils carries even more energy than he thought."

After concluding that marshes and sediments produced "flammable air," Volta began a series of experiments that highlighted his talents as a scientist and as a showman. He experimented with combining different proportions of regular air and sediment "air" with the purpose of showing which combination would create the loudest bang once ignited.

Through this type of investigation, Volta could satisfy his intellectual curiosity while also putting on an entertaining presentation that might make science interesting to non-scientists. One of Volta's continuous goals throughout his life was to popularize science and attract good people to this discipline. Toward this end, Volta would impress visitors to his laboratory by utilizing electrical equipment in a variety of attention-getting demonstrations.

Late in life, Volta especially enjoyed working on methane pistol prototypes. These later models were of sophisticated design, sometimes resembling actual pistols. In his demonstrations, he would create inflammable air (known now as hydrogen) by putting iron filings in contact with an acid. The size and explosiveness of the reaction would depend on the proportions he used.

Scientifically, Volta's discovery of methane fascinated his colleagues across Europe. Through his correspondence with top scientists—and the publication of his letters on combustible air—news spread quickly of Volta's discovery.

Within a decade, Lavoisier and other prominent scientists confirmed that their Lombard colleague had discovered what was dubbed "gas hidrogenium carbonatrum" (methane is still known as "light carbureted hydrogen"). The term "methane" for this chemical compound only entered the English-language lexicon during the 1860s, and the International Congress on Chemical Nomenclature confirmed the term for the scientific community in 1892. Moreover, it took quite some time for a full scientific understanding of Volta's discovery. "Nearly a century elapsed before firm evidence was obtained that methane formation in such habitats was a microbial process," notes Ralph S. Wolfe.

Although methane did not have many applications during Volta's lifetime, his groundbreaking discovery did contribute to a significant number of advances in the decades and centuries to come. One discovery leads to another, as the saying goes, and a rather straight line can be drawn between Volta's boat trips on Lake Maggiore and the natural gas commonly used in many facets of life today. That's why his findings on methane have been hailed as forming the beginning of the internal combustion engine.

To be sure, Volta's investigations of methane did not occur in isolation. The 1770s was a time of great activity in the discovery of gases and other elements. In 1772, the Scottish chemist Daniel Rutherford identified nitrogen, and Lavoisier's discovery of oxygen followed a year after Volta's isolated methane.

# Chapter Five

## VOLTA AND METEOROLOGY

### BIRTHING MODERN METEOROLOGY

Alessandro Volta was part of what Marco Ciardi, an expert on the history of science, calls "the birth of modern meteorology." Volta displayed an interest in meteorological matters for decades and helped to shape a field that during the eighteenth century was largely in its infancy. In the mid-1760s, for example, Volta presented a theory on electrical phenomena that was tied to atmospheric conditions. While not as well known or groundbreaking as his innovations with electricity or even methane, his work in the meteorological field was nonetheless important.

The roots of modern meteorology date back thousands of years. Over the centuries, Chinese, Greek, Indian, Arab, and Roman civilizations made contributions to what is now referred to as atmospheric science. But by the beginning of the eighteenth century, astrology and superstition still were the dominating influences in the way people understood weather. The 1700s marked a major transition period in the history of meteorology that set the stage for major meteorological advances that arrived in the nineteenth century.

Pervasive problems such as the lack of high-quality instruments and inadequate record keeping were addressed by the Enlightenment's emphasis on improving science technology and placing importance on detailed documentation. Moreover, the rise of natural philosophers resulted in more interest in natural topics such as weather. Governments, as part of their initiatives to develop science programs, also devoted resources to creating dependable weather observers. Consequently, at the dawn of the 1800s, a wealth of accurate weather data became available to scientists.

This time period was one of much interdisciplinary activity in the study of weather. Fewer lines were drawn between the natural sciences than exist today. In the eighteenth century and into the nineteenth century, the nascent study of meteorology overlapped with several disciplines, and intersected with work done with pneumatics (the study of gases), as well as with electricity and magnetism.

This cross-pollination of disciplines helped to feed the advancements made in meteorology. According to meteorology historian Jim Burton, "chemical theories of weather emerged alongside the identification of different gases by Joseph Black, Henry Cavendish, Joseph Priestley, Antoine Lavoisier (1743–1794), and others, while physical ideas about electrical influences followed the discoveries by Benjamin Franklin in America and Alessandro Volta . . . in Italy."

## INSPIRATION IN THE ALPS

When Volta embarked on his first trip abroad, the route he took with his companions, Francesco Venini and Giambattista Giovo, brought them across the Alps to reach the rest of Europe. Due the slowness of mountain travel in 1777, Volta would have had a good deal of time to marvel at his majestic surroundings. He could observe the alpine geology and natural elements as well as consider its history of earthquakes and volcanic incidents.

Volta's group traveled along the Gotthard Pass, also known as St. Gotthard Pass or Passo del San Gottardo, a key mountain path that connects one end of Switzerland to the other. On their route, they had to cross the Schöllen Gorge and take the Devil's Bridge (or Tuefelsbrücke) over the roiling Reuss river.

Volta and his companions may have been familiar with the traditional folk tale behind the construction of the Devil's Bridge. In this old legend, a Swiss herdsman asked for the Devil to build a bridge to make the arduous Reuss crossable. After the Devil built a bridge, the herdsman tricked the Devil. Riled over having a human deceive him, the Devil decided to demolish the bridge by smashing it with a rock. Before he could use the rock to destroy the bridge, an elderly woman marked the rock with a cross, preventing the Devil from picking it up. This was the fabled explanation of how the bridge came into existence.

Volta, Venini, and Giovo undoubtedly came close to the 220-ton rock that can be seen still near the Devil's Bridge today. These first-time travelers certainly must have enjoyed and marveled at the scenic vistas along the Gotthard Pass, which rises nearly 7,000 feet above sea level.

As a dedicated scientist and natural philosopher, Volta made observations unlike a typical traveler. He would pause on his trek to jot down meteorological observations. His actions are not surprising because the natural sciences, including meteorology and weather, had long appealed to him. This is evident, for example, in his interest in Franklin's experiments with lightning as well as with his own recent experiments in the Lake Maggiore marshes. Indeed, Volta would draw upon his research with electricity and methane to refine his thoughts on meteorological matters. "The interaction between electricity (both artificial and natural) and inflammable air had appeared to Volta as being the master-key to understanding many other phenomena that had not yet been adequately explained," writes Marco Ciardi.

Volta's seemingly trivial alpine observations in 1777 later factored into his story. They offer a notable example of Volta's interest in meteorology, which represents a small, but still significant, part of his scientific legacy.

## VASSALLI ENGAGES VOLTA

A larger and particularly revelatory record of Volta's perspective on meteorology comes from his multiyear correspondence with a young Piedmontese physicist named Anton Maria Vassalli (also known as Antonio Vassalli Eandi and Antonio Vassalli-Eandi) that occurred during the late 1780s and the early 1790s. Through their letters, Volta reveals a great deal about his approach to science and, more specifically, his progressive views on meteorology.

This correspondence, in fact, resulted from a meteorological event: the sighting of a large meteor. Just before 7:00 a.m. on September 11, 1784, Vassalli and his uncle, Giuseppe Antonio Eandi, also a physics professor, witnessed a strong, bright light moving across the morning skies over their home city of Turin.

The meteor was seen by people in Turin and the Savoy region, and in cities across Italy such as Asti, Susa, Genoa, Milan and Modena. Vassalli, then teaching philosophy at the Royal College of Tortona, received various reports about the meteor and aggregated the information for a report about this "bolide," a term for meteor, that he submitted to the Società Patria Letteraria in November 1784. He later wrote a paper on this topic, which the Royal Printing Office published in March 1786.

To Vassalli, it was clear that the nature of bolides was electric. This theory derived from his mentor and one-time physics teacher, Giovanni Beccaria, from whom he had learned about meteors. Several factors prompted Vassalli to this belief: the speed of the meteor's movement, the storm clouds present at the time, the thunder-like noise heard by those witnessing this event and the sky's otherwise apparent serenity.

After receiving positive and supportive feedback from several scientists, Vassalli sent his paper to Volta on July 18, 1786. He did so with good reason. Volta's stature was growing in the science community. Aside from being recognized as one of the top electrical science scholars, Volta was considered a leading authority in the field of meteorological studies.

Vassalli, who was just twenty-five years old, wanted to engage the forty-one-year-old Volta. It was a strategy that Volta knew very well—only not from the receiving side. Instead of being the one to reach out to an older, established expert, Volta was now the expert to whom this younger and less experienced man turned for help. Specifically, Volta's correspondence with Vassalli mirrored his earlier correspondence with Beccaria. Some twenty years after he courted Beccaria, Volta was now receiving letters from, and evaluating the work of, one of Beccaria's students. By engaging with Beccaria's protégé and former student, Volta was repaying Beccaria for their exchange of correspondence years before, during the 1760s.

Volta replied to Vassalli's first letter quite quickly. Whether the Beccaria connection influenced Volta's openness to write to Vassalli is not known. Another possible factor behind Volta's willingness to engage with his younger colleague was that Vassalli's uncle also enjoyed some regional prominence as an academic. It seems more likely that Volta's decision to reply to Vassalli's letter was because the young scientist's topic of correspondence—if not always his theories—had piqued Volta's interest.

Vassalli had already shown promise within the Italian science community at the time in the 1780s when he contacted Volta. Historian Mario Morselli placed Vassalli on the same level with Francesco Cigna and Giuseppe Saluzzo as part of the significant group of Piedmontese natural philosophers who came onto the scene in the late eighteenth century. Vassalli would go on to have a distinguished career—by the early 1800s, he would be the chair of physics, a job previously occupied by his uncle and Beccaria.

While the young Piedmontese scholar sought the older man's expert opinion and probably his approval, their correspondence bore fruit for Volta too. The exchange of ideas inspired Volta to hone his work on tools and devices to measure and document electricity.

Volta's letters to Vassalli provide fascinating glimpses into his thinking about scientific procedures, specifically in regard to meteorology. By this time, Volta had grown quite confident in his theories and thoughts on this subject. Starting with their first exchange in 1786, and continuing through subsequent letters,

Volta engaged Vassalli cordially and vigorously. While Volta stated that electricity might have some connection to aurora borealis, he strongly disputed Vassalli's theory regarding falling stars and the electric nature of meteors.

Although Volta never provided Vassalli with a definitive answer about the Turin meteor's unusual origins, he dismissed other theories, including one about atmospheric vapors. Furthermore, he discounted any hypotheses involving bolides and electric origins. Instead, according to Ciardi, Volta asserted that there was no way "an enormous quantity of electric fluid [could have gone] for many hundreds of miles through clear skies, instead of discharging itself into the ground."

Volta did admit that he once thought it possible for falling stars to generate significant electricity which, when ignited by inflammable air, might result in bolides and other larger meteors. But he was careful to make clear that this was more fanciful speculation than a substantiated conclusion. Still, Volta felt his theory about meteors was plausible, even though it raised as many questions as it answered. On the other hand, Volta expressed little patience for what he described as electricity's "fanatic worshippers"—people who saw a connection between electricity and just about everything under the sun, including earthquakes, volcanic eruptions, and even phosphoric sea light.

Over the years, Vassalli continued to turn to Volta for advice and shared his work with him. During the summer of 1791, he completed several works, and on October 24, he sent a number of them to Volta. One paper focused on his theories about how ancient people attracted lightning. Vassalli sought to show how ancient people were familiar with techniques of controlling lightning. To make his case, Vassalli used a variety of classical sources, including the writings of the Greek historian Herodotus. The young professor asserted that these discoveries had been lost over the centuries but had recently been rediscovered.

In his response, Volta was both complimentary and critical. While praising Vassalli's dissertation as scholarly and engaging, Volta remained wholly unconvinced by his theory, which he described as "an erudite joke." Ciardi notes that Volta was particularly skeptical of how Vassalli interpreted "the obscure sayings of the Oracles, in facts and mysterious expressions, in practices and ceremonies introduced by multiform pagan Superstition and in portentous Mythology." A staunch believer in empiricism, Volta countered that he found even the limited and largely ineffectual research of their time far preferable to merely theorizing over some myths and fairy tales from antiquity.

Volta was quite dismayed at the state of meteorological study in the early 1790s. Even though survey tools had evolved and become more refined than they had been even a decade earlier, Volta found that too much of the work regarding weather remained unscientific. In Volta's view, popular meteorology of the eighteenth-century variety was still informed by astrology, peasant life, cultural traditions, and alleged scientific knowledge handed down from the ancients. While he praised the many diligent weather enthusiasts who had conducted electric-meteorological observations, Volta described the research in metrological science as being highly flawed and called for the institution of more principles and core elements. He was consistently critical of those who favored history and beliefs over hands-on experiments and instruments that provided reliable, quantitative information.

Additionally, Volta deplored the state of Italian meteorology. Although he acknowledged that plenty of meteorological work was being done in Italy at that time, he was critical that there were not enough Italian meteorological observatories equipped with the first-rate barometers and thermometers necessary to perform organized recordings of such vital weather statistics as barometric average height and the annual and monthly average heat. Only the observatory in Padua met Volta's standards for its well-recorded observations. Conversely, major cities such as Milan, Turin, Verona, and Vicenza had been carrying out, and publishing, meteorological observations for years, yet their monthly and yearly data were flawed and insufficient in Volta's eyes.

Volta was especially galled when he contrasted the inadequate situation that he found in Italy with the impressive meteorology work in other countries. He singled out work by the Irish chemist Richard Kirwan, offering high praise for the tremendous diligence and attention Kirwan used in collecting his data. Kirwan was able to gather thermometric data with great accuracy for his time and consequently was able to formulate average temperatures for more than forty locations in England, Holland, France, Germany, Sweden, and Russia, as well as in Asia, Africa, and America.

It was Volta's wish to see meteorological observatories built in Milan, Pavia, and Mantua. He wanted more than simply operational buildings—he thought they should follow meticulous, uniform guidelines in observation times, atmospheric measurements, and weather data tabulations. Setting high standards for work was always important to Volta, and he felt that meteorology could offer value to physics, agriculture, medicine, and nautical science if it could improve its gathering of scientific data.

By 1792, despite his strong beliefs about meteorology, Volta's attention became consumed with his debates on electricity with Luigi Galvani, and his focus shifted away from weather and zeroed in on electricity.

Historians such as Ciardi assert that the late eighteenth century wasn't the right time for Volta to pursue theories on meteorology because it had not progressed as a proper science. Nevertheless, Volta's meteorology work integrated with his better-known research on methane and electricity, and he brought his high standards in testing and empiricism to every endeavor.

## TRAILBLAZING

Somewhat ironically, Volta's research on electricity and his invention of the Voltaic pile were ultimately his greatest contributions to the study of meteorology. This is because advancements in electricity that were inspired by Volta's work resulted in the invention of the telegraph, which consequently proved to be vital in meteorology and, more specifically, weather forecasting.

One consequence of the invention of the telegraph in the mid-1800s was that meteorological data could be collected across a continent almost immediately. The vast improvement and expansion of weather data had multiple effects, such as enhancing the growth of agricultural and ocean-based commerce. The benefits achieved through the use of this information encouraged countries to create national weather services. The formation of these governmental agencies led to the establishment of a network of meteorological observations that could coordinate forecasting for dangerous weather such as storms. The explosion of meteorological data also helped to produce effective and comprehensive weather charts.

Volta did not live to see the evolution of meteorology as it developed into a full-blown scientific discipline. It was the nineteenth century that proved to be the right time for meteorology to take off. The scientific predicting of weather was in its infancy in 1800, but by the 1860s, meteorologists had already issued the first weather forecasts and storm warnings.

It would have delighted Volta that training in physics became essential for meteorologists. From a twenty-first-century vantage point, Volta's work on meteorology was elementary— almost laughably so—but it still served important purposes. This includes his collection of meteorological data in Gotthard Pass, his advancements in the field of electricity, and his promotion of establishing rigorous scientific standards.

Today, meteorological science plays a significant role in our lives. The great convenience of forecasts and the detailed information that they provide are the result of the theories and inventions from scientists such as Volta, his contemporaries, and hundreds of scientists who came after them.

# Chapter Six

## ON THE ROAD TOWARD THE VOLTAIC PILE

### THE SUPREME ELECTRICIAN

Electrical experimentation was quite in vogue during the eighteenth century. Few areas of scientific research stirred as much activity as the study of electricity did during this era. Contributions that resulted from the advances and inventions starting from the 1700s are still important today.

Alessandro Volta was a major player behind the great progress made in the field of electricity in that century. During the twenty-five-year span between 1775 and 1800, Volta experienced a highly productive period—he was formulating new theories, making discoveries, and constructing instruments and apparatuses that revolved around electrical science.

His invention of the electrophorus in 1775 and his discovery of how to isolate methane the following year represent just two of the major scientific contributions that he made on the road to creating the first electric battery. A registry of Volta's devices that he created or improved upon include items such as the eudiometer, the electroscope, the condensatore, and the flashy sparking tool known as Volta's Pistol or the Inflammable Air Pistol. Volta also came up with his Contact Theory of Electricity and its related List of Conductors.

These innovations served important purposes during the watershed era of electricity experimentation and played roles in Volta's construction of the "Voltaic pile," which is universally acknowledged as the first electric battery.

Volta's experiments with methane arose out of his interest in "inflammable air," as hydrogen was referred to back then. British chemist experimenter Henry Cavendish, the man who identified hydrogen in 1766, used the term "inflammable air" until Antoine Lavoisier named the element "hydrogen" in 1783.

In the mid-1770s, the common instrument used to trap and measure various gases was the eudiometer, which could quantify how much oxygen was in the air to determine if it was healthy to breathe. By 1777, Volta had created a eudiometer that vastly improved upon earlier versions, principally because it provided highly reliable reactions. His eudiometer could precisely mix hydrogen with oxygen due to the fact that the volume of hydrogen gas decreased after sparking, because the "inflammable air" reacted with oxygen to make water. The decrease in the volume of hydrogen gas was proportional to the amount of oxygen present in the air.

Moreover, it was through his eudiometric experiments with methane and oxygen that Volta discovered an electrical spark would cause this mixture to explode. This reaction solved the problem common to earlier eudiometers that didn't have reliable ways to make the gas-oxygen mixture interact. To create this electrical spark, Volta constructed a device that has been called "Volta's Pistol," or, since inflammable airs were involved, the "Inflammable Air Pistol."

It has been documented that Volta's Pistol could launch a lead ball several meters. So it is somewhat surprising that the "pistol" never evolved into use as a military weapon. While the practical functions of eudiometers had been almost totally eclipsed by more effective scientific inventions as early as the early nineteenth century, Volta's Pistol remained in use, particularly in classrooms, because the loud popping sound it created was an attention-grabbing teaching aid for professors.

Volta's Pistol led to more innovative ideas. In a 1777 letter to the University of Pavia physics professor Carlo Barletti, Volta shared his concept for an electrical signal line that could run the nearly thirty miles from Como to Milan. Utilizing the electric pistol to ignite an electrophorus, Volta's "electrical telegraphy" could send messages through insulated wire. This innovative communication proposal later would become realized as the telegraph, one of the most revolutionizing devices of the nineteenth century.

Links can be made from Volta's use of an electric spark to create a controlled explosion to the invention of other essential modern devices, from the internal combustion engine to the gas lighter. Methane, furthermore, would also play a major role during the Industrial Revolution as it helped to fuel gas turbines and steam boilers.

Volta's ongoing exploration into the nature and behavior of gases led to landmark discoveries. He did pioneering research, for example, into how heat could affect the expansion of gases. His observations presaged the now-famous Charles' Law—articulated by the French scientist Jacques Charles in 1787—which asserts that all gases expand equally for equal degrees of heat.

One of Volta's well-known characteristics was his relentless attempt to improve his own inventions, and this habit turned out dividends. While exploring ways to upgrade his electrophorus machine, Volta started experiments using conductors and their electrostatic capacities.

This line of scientific inquiry led, in the late 1770s, to Volta coming up with an instrument that he called the condensatore. Particularly useful in investigating "weak" atmospheric electricity, the condensatore was a device that Pancaldi describes as making "detectable to ordinary electroscopes otherwise undetectable amounts of electricity."

One purpose that Volta had for this instrument was to use it during an aurora borealis event to detect atmospheric electricity. As a result of inventing this electrical condenser (comparable to the modern day capacitor), Volta discovered that an electrical charge is directly proportional to the electrical potential—what would now be called "voltage."

In the 1780s, Volta's investigations into electricity led him to the idea that there were electrical effects when water evaporated—this idea arose before it was known to be due to the friction of the vapor. Because he was determined to have precise scientific measurements on water evaporation, Volta worked doggedly to make significant improvements to the electroscope, also referred to as an "electrometer," which was then the primary device used to measure electrical charges.

One enhancement he made around 1787–1788 was to introduce a flame to the electrometer; this allowed scientists to more quickly gauge electrical tension in the atmosphere. Perhaps more important, Volta swapped the metal wires in the electroscope for straw. This substitution had several major positive ramifications. There was the practical advantage of straw being a far less expensive material than what was commonly utilized then—typically silver wire or gold leaf. On a scientific level, straw made the electroscope more sensitive in its readings. This resulted in more detailed and standard measurements, which was one of Volta's main goals in his work to refine scientific instruments.

Volta had two purposes driving this goal. One was to make sure that there would be uniformity in the data measurements of every electrical degree taken by the same instrument. His second reason was to institute a structure so that accurate comparisons of measurements could be made by electricians. One could hypothesize that by inventing instruments that served to standardize measurements, Volta gained legitimacy for his own measurements in the eyes of his fellow scientists.

It was through his experiments in atmospheric electricity that Volta also discovered how to get an accurate measurement for the increase in air volume in relation to rising temperatures. Later, in the 1790s, he found that a liquid's vapor pressure was totally dependent on temperature and not atmospheric pressure, as had been thought.

The 1790s also saw one of Volta's greatest achievements prior to his invention of the battery—the List of Conductors. He was the first person to compile a lengthy list of metals in the order of their electromotive force, from strongest to weakest voltage.

## VOLTA'S LIST OF CONDUCTORS

| | | |
|---|---|---|
| Zinc | Iron | Gold |
| Lead | Copper | Graphite |
| Tin | Silver | Manganese Ore |

To compile an accurate list, Volta conducted numerous experiments that matched pairs of these metals so he could see what type of electromotive force, today called "voltage" in Volta's honor, that they would create together. Through this hands-on process, he discovered that a coupling, or "cell," of a "strong" metal with a "weak" one created a greater electromotive force than would two "strong" metals paired together. As an example, a "cell" of zinc and gold generated more voltage than one composed of zinc and tin. This process resembles the current measurement called "standard electrode potential."

During the course of his research, Volta developed separate ways to study electrical potential and charge. Furthermore, he discovered that the electrical potential is directly proportional to the electrical charge in a capacitor—a device with one or more pairs of conductors that are used to store an electric charge. Originally named as Volta's Law of Capacitance, this theorem is commonly known now as "electrical capacitance."

The List of Conductors grew out of Volta's "Contact Theory of Electricity." He conceived this theory after noticing that two metals connected by a type of moist material would produce an electrical charge. He also discovered that the moist material did not have to be animal in nature. This distinction was significant because Volta was largely investigating the claims of scientist Luigi Galvani regarding "animal electricity."

Volta divided electrical conductors into two types. One type contained metals, graphite, and pure charcoal, while the other contained substances such as salt, water, or dilute acids that now are classified as electrolytes. When he connected two conductors from the first kind with one from the second, a circuit was created that resulted in an electric current. This phenomenon forms the basis of his Contact Theory of Electricity.

The Contact Theory demonstrates another instance in which Volta successfully built upon an older idea, in this case his theory of "Chemical Electricity." He accurately deduced that if metals were brought into contact with each other, they would create an electrical current through a chemical reaction. Prevailing theories held that static electricity was generated just through friction. Volta's concept of chemical electricity has been hailed as one factor in the way cities around the globe were eventually transformed by technology.

# Chapter Seven

## GALVANI'S FROGS

### A SCIENTIST AND HIS FROGS

The science community experienced a seismic jolt in 1791 upon the publication of groundbreaking findings by an Italian scientist. The man was not Alessandro Volta, although his legacy is inexorably intertwined with Volta's story. He was Luigi Galvani, and his historic paper concerned his theory of "animal electricity."

In his essay, entitled "*De Viribus Electricitatis in Motu Musculari Commentarius*" or "Commentary on the Effect of Electricity on Muscular Motion," Galvani proclaimed his educated belief that animal tissue contained an "animal electricity" which was different from natural electricity (such as lightning) or even artificial electricity (such as static electricity caused by friction). He further proposed that the brain created an "electrical fluid" that traveled via nerves to muscles. These ideas were revolutionary for their time, since natural electricity and artificial electricity were then the established and accepted theories of electricity.

Aloisio Luigi Galvani was a doctor and a professor of anatomy at the University of Bologna. Prominent in his city's science community, Galvani was selected president of the Academy of Sciences of Bologna in 1772. He was already in his early forties when he published his now-famous paper. He actually had written several essays about his electrophysiological work during the 1780s, but inexplicably chose not to publish them. Even though he was an esteemed, long-time member of the university's faculty, Galvani preferred to do the majority of his scientific research at his house.

Galvani's research originally centered around comparative anatomy (the subject of his doctoral thesis in 1762) and physiological movement in small animals; frogs were among his primary subjects. In the late 1770s, his interests started gravitating toward the study of electrophysiology. Within a few years, Galvani's research shifted more toward the effects of electrical stimulation on animals' musculature, and he acquired a Leyden jar and an electrostatic machine to conduct his experiments. He arrived at this groundbreaking theory when he noticed that a frog's muscles would react when touched by an electrical stimulus.

For many years, the origin story behind Galvani's theory about electricity and animal muscles involved frog soup. The explanation starts with Galvani's wife, who was sick with consumption, asking their servants to make her frog soup. While a servant was preparing a frog, a spark from one of Dr. Galvani's nearby electrical machines touched his knife, and this contact made the frog's leg twitch. When told of this accident, Galvani became highly intrigued and started his investigation.

Although quite entertaining and colorful, the story turned out to be apocryphal. It appears that Galvani was doing experiments on the relationships between nerves and muscles. He observed that when a frog's nerve was connected to a source of electricity, the attached muscle would contract.

While the idea that animal muscles possess an electrical force might seem far-fetched now, it is worth remembering that there was no wealth of knowledge concerning animal physiology in the eighteenth century. In fact, at the start of the 1700s, the prevailing theories about the nervous system were still based on the concept of the "animal spirit," which had been conceived by the influential Greek philosopher and physician Galen, who lived in the second century.

Galvani's "animal electricity" theory, consequently, provided a more empirical, enlightened explanation concerning how the nervous system worked than Galen's speculative "animal spirit" notion. Utilizing thorough scientific methodology, Galvani spent many years doing extensive research and conducting experiments in all types of conditions to test muscular reactions to electricity. In the course of his research, he used warm-blooded animals, as well as cold-blooded ones like frogs. He also tried different sources for electricity, including a Leyden jar and a rotating electrical machine.

Many experiments were successful, while others were not. Through additional experimentation, he extrapolated that the muscle could create, and discharge, an electrical current. Another significant observation was that a frog's muscles contracted after touching its nerves with a pair of scissors during an electrical storm; another was that a dead frog's legs kicked when its lumbar nerve was touched by a scalpel connected to an electrical machine.

Galvani's major revelation, however, came when he noticed that convulsive movements in a frog occurred even when there was no source of electricity—that this reaction took place even when he only connected a brass hook to a frog's spinal column and hung the hook on an iron railing. From this experiment, Galvani arrived at his landmark theory that "two dissimilar metals making contact with a nerve attached to a muscle can make that muscle contract."

"Galvani visualized a living creature as functioning something like a fleshy kind of Leyden jar," Bern Debner explains in his book, *Alessandro Volta and the Electric Battery*. "The nerve and the muscle would constitute the inner and outer charged surfaces of the jar. When the outer surface of the muscle received an electrical charge (like the outer surface of a Leyden jar), the nerve and inner muscular surface would then become oppositely charged and muscular contraction would follow."

## THE REACTION TO ANIMAL ELECTRICITY

When Galvani published his article in 1791 in the *Proceedings of the Bologna Academy of Sciences*, it created a sensation because his theory finally offered an explanation for why dead tissue could quiver as if it were alive. Galvani's essay launched major debates among physicians, scientists, and philosophers. What particularly commanded the attention and respect of his colleagues was his methodology. Galvani was quite meticulous in his planning and execution of his experiments. By performing a large number of experiments, he accumulated an impressive amount of data for a scientific sampling, and then analyzed this data thoroughly. In his paper, Galvani not only recounted his experiments in chronological order, but also provided logical accounts of how he reached his conclusions.

In addition to publishing the essay, Galvani printed twelve copies of his pioneering work and distributed them privately. Volta was one of the dozen people chosen to receive a special copy with the inscription "From the Author." The two were good friends as well as scholarly colleagues.

Given Galvani's methodical approach to his experiments, it is easy to see why Volta would find his friend's essay highly appealing. If fact, Volta was so intrigued by Galvani's experiments that he set out to repeat them on his own. At first, Galvani's work impressed Volta, who stated that his colleague's essay "contains one of the most beautiful and most surprising discoveries." But despite these initial compliments, Volta was not totally convinced by Galvani's theory.

Volta's concerns partly arose out of their different scientific approaches. As a doctor, Galvani took a predominately anatomical and physiological perspective with his experiments, while Volta held a viewpoint that favored physics and chemistry. His opinion of Galvani's theories, however, was more influenced by his own hands-on work. The more he attempted to replicate Galvani's experiments, the less Volta believed in the conclusions of the University of Bologna professor.

When Volta presented a careful analysis of Galvani's ideas on "animal electricity" at the University of Pavia, he examined Galvani's theory and proceeded to challenge much of it. Volta felt that Galvani had not proven that intrinsic animal electricity really existed, but instead had only shown that electricity could be a powerful stimulus on muscles and nerves. Drawing his insights upon his own experiments involving his sense of feeling, sight, and taste, Volta countered with the suggestion that investigating the metallic arc—the piece of metal that Galvani used to connect a frog's leg and spinal cord—would be a better way to find the source of electricity.

The controversy surrounding Galvani's theories ignited an intellectual battle within the Europe's scientific community, pitting Volta's theory of "metallic electricity" against Galvani's "animal electricity." Scientists lined up on either the pro-Volta side or the pro-Galvani side, although some straddled the fence and found much to admire in the work of both scientists.

Giuliano Pancaldi describes the situation this way:

*Until July 1796, Volta measured weak electricity via frogs and other physiological apparatus, such as his own tongue and sense organs. This was consistent with Galvani's own assumption that in such experiments frogs acted as 'most exquisite electrometers.' According to Galvani and his followers, however, frogs did more than that: they provided the electric fluid, supposedly proper to all animals, that caused the contraction of their legs. Against this conception, Volta had maintained consistently—after a short-lived adhesion to Galvani's views—that the frog played only a passive role, proper to an instrument for measuring electricity. According to Volta, the motion of the electric fluid evidenced by the contractions of the frog's legs arose from the contact of different metals and wet substances employed in experiments on galvanism.*

During the 1790s, many treaties were written, papers published, presentations done, and letters exchanged, all making the case for one camp over the other. Volta himself addressed the brewing controversy in an article entitled "Account of some discoveries made by Mr. Galvani, with Experiments and Observations on them," that was published in the Royal Society's *Transactions* in October 1793. Between 1793 and 1800, in fact, Volta was said to have authored some twenty essays and numerous letters on this topic.

The publication of Galvani's "*Memorie sull'elettricità animale*" or "Memories on animal electricity," in 1797 intensified the debate as it followed an anonymously published 1794 essay entitled "*Dell'uso e dell'attivitu dell' Arco conduttore*," which addressed the activity and applications of the "conducting arc" in the contraction of muscles.

In both instances, Galvani and his supporters aggressively challenged Volta's viewpoint in the galvanic controversy. Volta responded quite emphatically. In April 1798, he restated his past arguments—and doing so more systematically than he had previously—in two anonymous letters to Galvani's nephew, Giovanni Aldini, who was also a University of Bologna physics professor. This Galvani-Volta debate, which author Debner describes as "one of the most famous contests in the history of science," engaged scientists and other scholars for nearly a decade, and often devolved into an intellectual brawl.

Even while Volta and Galvani were at the center of this ongoing scientific showdown, they did not seem to turn their friendship into a bitter rivalry. They behaved as gentlemen with each other during this philosophical debate (and not every scientist did), demonstrating how highly principled both men were. Volta, in fact, used the term "galvanism" to explain the type of experiments in which contacts between "first-class conductors" (metals) and "second-class conductors" (humid matters) caused a reaction in electric fluid.

## THE TASTE OF METAL

Volta was particularly skeptical of the idea regarding "animal electricity" and Galvani's comparison of an animal's muscle with the Leyden jar. As he conducted his own experiments, Volta came to view metal strips as the generators of electricity rather than the animal nervous system.

To investigate these ideas, Volta used his own body in his experiments. In one test, he took two strips of metal and placed them on either side of tongue, resulting in the experience of an unpleasant taste in his mouth. Continuing along this line of inquiry, Volta performed several variations of this experiment. He put a metal strip next to one eye, which caused him to feel the sensation of light near that eye. These results further increased Volta's belief that metals were the source of electricity.

These tests also represent another example of Volta's inadvertent, and unknowing, duplication of another scientist's experiments. In this case, it was the work of the Berlin-based Swiss physicist Johann Sulzer, who performed the metal-strip-on-the-tongue experiments back in 1752. As on the other occasions, Volta eventually went beyond Sulzer's experiments on his eyes by using other items such as coins and silver spoons in his tests.

## THE SPIRIT OF 1796

When Volta, as an old man, looked back upon his life, he picked 1796 as the turning-point year for his work that led to his invention of the battery. This choice is not surprising. It was in the summer of 1796, after all, when Volta's focus shifted from investigating "animal electricity" to exploring his own "metallic electricity" theory.

A key impetus behind this change came from Galvani and his scientist supporters. In the year 1796 they demonstrated the ability to produce contractions in a frog's body without using metal. This turn of events punctured a hole in Volta's belief that metal was the reason for the electrical charge, not animals.

Volta reacted to Galvani's revelation in two ways. First, he admitted that contact between different humid conductors, such as an animal body, could stimulate electric fluid even without metal being involved. Second, Volta became determined to prove that animals didn't possess a unique type of electricity in order to quash the need to use animals as electricity detectors.

Blocking Volta from this goal was the fact that the electrometers he used could not produce the sensitive measurements that a frog could. For help with this technical predicament, Volta turned to his instrument maker at the University of Pavia, Giuseppe Re—the same man who accompanied him on the 1782 trip to Germany—to create a special version of a device called the "doubler" that British scientist William Nicholson had described in a 1788 issue of *Philosophical Transactions*. Volta believed that an electrometer could detect weak electricity, on the level that a frog could, with enhancements incorporated into the doubler apparatus. Volta also felt that the doubler, in fact, was based on his own invention, the condensatore.

Within a few weeks of receiving his special doubler-improved electrometer, Volta could detect the electricity created from contact between two different metals, which had only been achieved previously by a man's tongue and a frog's legs. Volta's results meant that electricians could now use ordinary means and criteria to test for even the weakest types of electricity.

Through this research, Volta began favoring the idea that two different metals worked better as a "motor" for electricity whereas earlier he had focused on a connection between first- and second-class conductors. Furthermore, the doubler device allowed Volta to use metal discs in experiments, which factored into his later work on Voltaic piles.

The Galvani-Volta debate did not resolve with a clear-cut winner. The two men were each right about certain matters, and wrong about others. The results of some experiments actually brought Volta around to agreeing with a few of Galvani's ideas over the course of time. In a letter he wrote in 1798, Volta admitted that animal electricity does exist in some living organisms while contact electricity also exists.

For his part, Galvani can be viewed as a catalyst who influenced Volta's invention of the battery. Sadly, however, he never lived to see this happen. The highly principled professor lost his job and salary after refusing to state allegiance to the new Napoleon-backed government that had gained control of northern Italy in 1797. While he was eventually offered his job back, the traumatic situation took a severe toll on Galvani's health. He passed away on December 4, 1798, just fifteen months before his friend, colleague, and critic announced the Voltaic pile to the world.

## LIVING LEGACY

The ongoing scholarly disagreement between Luigi Galvani and Alessandro Volta, as well as each man's own innovative research, certainly accelerated an interest in electrical studies that grew in significance during the nineteenth century. It is easy to see how Volta's invention of the electric battery spawned an explosion of new inventions almost immediately, and has continued to play an essential part of modern life.

From today's perspective, both men were winners in their scientific debate because each one has had an influence on the entire world—in the short- as well as the long-term. One undisputable consequence of the work of these two men was that life science broke free from the ancient "animal spirit"-based theories and opened the field of study to the modern age.

While recognition for Galvani's accomplishments might not be as prominent or universal as Volta's, his experiments proved to be instrumental in developing the field of electrophysiology, which stands as one of the foundations of modern science, particularly the neurosciences. Even Galvani's theory that animal tissue had the ability to generate an electrical current was shown not to be wrong; however, it took nearly a century for this to happen. In 1879, the English physiologist John Scott Burdon-Sanderson demonstrated that a beating heart did create electrical impulses.

Galvani's more immediate legacy in the 1800s was a bit more macabre. His electrical experiments with body parts caused some people to think that he had discovered a secret to life. Galvani's nephew Giovanni Aldini, who was a particularly vocal booster of "animal electricity," experimented with electrifying the human body and frequently performed public demonstrations. One of Aldini's best known performances took place at London's Royal College of Surgeons in 1803, when he used a body of a convicted murderer that he had brought from the gallows. Attempts at using electricity to bring the dead back to life continued throughout the nineteenth century, and some believe that these types of galvanic experiments influenced the English author Mary Shelley to write her novel *Frankenstein*, published in 1808.

# Chapter Eight

## DISCOVERING THE VOLTAIC PILE

### A TORPEDO FISH POINTS THE WAY

During the latter part of the 1790s, Alessandro Volta embarked on the significant scientific research that climaxed in 1799 with his creation of the Voltaic pile, commonly known today as the electric battery. Considering the historic nature of this invention, it is somewhat surprising that historical evidence suggests that the battery was not a goal for Volta until the latter part of 1799. In his Volta biography, Pancaldi examined archival material in-depth, as well as Volta's unpublished writings and laboratory notebooks, in order to create a comprehensive reconstruction of Volta's process in constructing the Voltaic pile. Pancaldi's conclusion is that ". . . until [the] spring [of] 1799, and probably even later, Volta was *not* in any sense engaged in a search for a device like the battery."

While there is much truth in Pancaldi's analysis, you could also say that Volta, in a broader sense, had been leading up to making a transformational invention like the battery for his entire life. It can be seen in the inventions and discoveries—such as the electrometer and the List of Conductors—which he created in the years leading up to the Voltaic pile. Three-and-a-half decades of work in the realm of electricity combined with the guiding elements of his personality—ambition, drive, curiosity, and his quantifying spirit—into a driving force that culminated in his monumental discovery. It is all the more disappointing then that Volta's autobiographical record from this period is spotty; however, Pancaldi suggests that the Voltaic pile was built in late summer or autumn of 1799.

One crucial turning point that directed Volta on the path towards inventing the Voltaic pile was a paper written by William Nicholson, the same English scientist whose "doubler" device influenced Volta earlier in the decade. The two men were quite familiar with each other's work as they traveled in the same circles among the elite electricians in Europe. In the November 1797 issue of the *Journal of Natural Philosophy, Chemistry and the Arts*, Nicholson published his "Observations on the Electrophorus, Tending to Explain the Means by Which the Torpedo and Other Fish Communicate the Electric Shock," which dealt with the topics of electric fish, detecting weak electricity, and constructing creative electrical devices. The article struck a chord in Volta. He was particularly taken by Nicholson's proposal for constructing a device that imitated the electric organ found in a torpedo fish.

The torpedo fish hails from the same order of species (*Torpediniformesas*) as electric eels and electric rays, and has long fascinated those interested in the natural sciences. In ancient Greece, torpedo fish were used to stun people for either medicinal or punitive reasons. Its name comes from the Latin word, "torpere," which means to be stiffened or paralyzed, and it is also how the modern torpedo weapon got its name.

Volta had become acquainted with the torpedo fish's electrical capacities when he traveled to London in 1782. During this trip, he met with John Walsh, England's expert on the electric fish, and he came away with the idea that this fish represented the only true instance where electricity could be found in an animal. It was on this English visit that Volta became aware of the notable paper penned by Henry Cavendish (the British chemist, physicist, and natural scientist who discovered the inflammable air that is known now as hydrogen.) His paper discussed the idea of an artificial torpedo created from a large number of Ledyen jars.

Torpedo fish resurfaced on Volta's radar in the late 1790s through the work of Galvani. The University of Bologna professor had a long interest in the electric fish, and had returned to studying this creature around 1797. Through his experiments, Galvani reported that the fish's electricity came from its brain and nerves. His writings on these experiments were convincing enough for Volta to agree with his rival's points, as Volta noted in letters written in the fall of 1798.

It is easy to see, consequently, how Volta was already thinking along the same lines as Nicholson when he read the English scientist's essay, which suggested building a device that could emulate the torpedo fish electric organ and that this might be the best way to imitate the electricity that the fish generates. Nicholson's idea was the catalyst that steered Volta toward building the Voltaic pile. It provided the basic concept and thoughts on its general construction. While Volta admired, and adhered to, a good deal of his colleague's ideas, he also vigorously disagreed with other parts, leading him to significantly modify Nicholson's model at critical points.

Volta believed Nicholson was especially misguided in one area: his idea that an electrophorus would work well as a conductor. The English scientist thought that the fish's internal insulating parts were comparable to the electrophorus' resin cakes. A better choice, in Volta's reasoning, was using pairs of self-acting conductors, such as metallic couples. He knew that metallic couples were more effective than resin cakes in getting electricity into motion, and he could create a fully electric device without the additional mechanical apparatus to make the electrophorus operate properly.

Another technical flaw that Volta found in Nicholson's model was that its arrangement of metal plates and conductors would not be conducive to generating enough flow of electrical fluid. Volta's challenge, consequently, was to design his device in such a way that it could successfully create a suitable amount of electrical fluid flow. His solution was, as Volta wrote when recalling his invention of the battery, that "I conceived the idea of connecting many [couples of different metals] together."

## FROM A CROWN OF CUPS TO A VOLTAIC PILE

Although Volta did not leave a step-by-step journal of how he created the Voltaic pile, the evolution of his battery "prototypes" can be roughly traced. Through his experiments, Volta knew that an electrical charge was produced when two dissimilar metals came into contact, and that the electrical charge was even greater when a moist element was involved.

An important early experiment that proved this point involved placing two strips of dissimilar metals—typically one was copper and the other zinc—into wine goblets, which were filled with a brine or saline solution. Volta then connected a metal strip in one goblet with its opposite metal strip in another goblet. A steady electrical current was produced through the connection of these metal strip "electrodes" in the brine, and created a battery-like effect.

Because the goblets were placed in a circle, this experiment was dubbed "The Crown of Cups." The invention of the Crown of Cups battery was a milestone because it replaced static electricity with an electric current flow that could be utilized in multiple ways. As ground-breaking as this invention was, the Crown of Cups came with several inherent shortcomings; namely it took up a lot of space and was difficult to use.

Looking for an alternative way to stimulate an electrical current with two metals, Volta conceived of the idea to stack the metal strips in piles with a moist conductor in between them. Placing the metal in piles was a revolutionary idea, especially at that time. The prevailing theory held among galvanic experimenters (a group in which Volta could be included) was that conductors needed to be in a circle or an arc.

Volta's inspiration might have derived from the column-like body structure of a torpedo fish. Indeed, one of the early names that he is said to have considered for his invention was "artificial electric organ," to contrast it with the torpedo fish's "natural electric organ."

Another possible influence for his metal pile idea might have been the well-known Prussian polymath Alexander von Humboldt, who had recently published his observations that discussed how a frog could be stimulated when placed side-by-side to two metals—and the metals did not have to be in a circle or an arc.

Regardless of the impetus behind his thinking, Volta came up with the idea to take two dissimilar metal discs and stack them in an alternating fashion, separating the discs with paper or cloth, soaked in water or some type of saltwater brine. He then attached a wire to the top and bottom metal discs, which created a continuous electrical current far more efficiently than did the Crown of Cups.

This device, known as the Voltaic pile, has been credited as the first electric battery. To describe what Volta did using contemporary chemistry terms, he used two electrodes (the copper and zinc metal discs) and an electrolyte (the wet paper or cloth). When the electrodes interacted with the electrolyte, an electrical current was created, the copper becoming a positive electrode and the zinc a negative.

Volta is credited with articulating the law that the electromotive force (emf) of an electrochemical cell, such as a battery, is the difference between the potential of its two electrodes. This rule has been dubbed "Volta's Law of the Electrochemical Series," and in his honor, a unit of electromotive force is called a "volt."

Ever the inventor, Volta continued to tinker with other concepts for the pile design. The biographer Pancaldi reveals that Volta's laboratory notes contain plans for a non-metal battery constructed out of discs of bones that have been soaked in potash, sulfuric acid, and fresh water. This battery model did not meet with success. Nor did another idea that involved encasing a column made of twenty dissimilar metals in a material such as wax with lead wires at either end of the column that could be used like handles.

Through his life, Volta worked on many experiments, but his most triumphant idea of the Voltaic pile earned him immortality for the far-reaching impact that the battery has had on civilization.

# Chapter Nine

## THE BATTERY ARRIVES

The electric battery had a major impact on the field of communications. So, it is a bit ironic that Alessandro Volta faced many challenges getting the word out about his invention. It was still decades before Samuel Morse would create his telegraph and Alexander Graham Bell did not invent the telephone until 1876.

Much as he did as a teen when he wanted to contact notable scientists, Volta used letter writing to share the news of the Voltaic pile. This time, however, he took a much more sophisticated approach. Just as he was methodical in his experiments, the Lombard scientist carefully developed his presentation strategy for his battery too. Launching what amounted to a multi-country campaign to publicize his battery, Volta proved to be a deft and shrewd promoter of his historic invention.

At fifty-five years old, Volta was a veteran of the clashing egos, intellectual debates, and vigorous competition for ideas that characterized his international community of electricians.

Based upon his prior experiences in announcing inventions, he knew it was important to carefully consider the "who" and the "where" when it came to revealing his new discovery.

## EUROPE'S POLITICAL CLIMATE IN 1800

A central concern for Volta was politics. Although he was focused on his work as a scientist and scholar, he could not ignore the political world around him. In an effort to acquire patrons early in his career, he made a point of meeting powerful political figures around Europe. The long relationship he developed and maintained with the Austrian government proved beneficial to both parties.

With the dawn of 1800, Volta needed to utilize his political savvy in selecting a nation for the announcement of his invention. He had four choices for suitable countries—Italy, France, Austria, and England—and each had positive and negative attributes.

During the 1790s, while Volta and Galvani battled over revolutionary theories of electricity, an historic revolution of the political kind was taking place in France. In 1792, King Louis XVI was overthrown, resulting in the establishment of the first French Republic. Almost immediately, the new government was at war with most of Europe. The War of the First Coalition (1793–1797) found the French Republic waging war against the monarchies in Spain, Austria, England, Holland, Sardinia, and Prussia.

In 1796, a young general named Napoleon Bonaparte was put in charge of the French forces in Italy. His army was triumphant throughout their Italian battles, and his victories continued as they moved north into Austria. Napoleon's military successes raised his popularity at home, and he became more and more involved in French politics.

In 1799 and 1800, however, France was going through another political upheaval. The governing body known as the Directory, which took over for the First Republic in 1795, was overthrown in November of 1799. Napoleon was elected head of the government at the end of the year, but in 1800 he was still entrenched in fighting both the Austrians and the British. After the Austrians surrendered in February 1801, Napoleon's power at home grew and strengthened even more when he was overwhelmingly re-elected to lead France following the signing of the Treaty of Amiens with the British in March 1802.

Napoleon met Volta in 1796, although accounts differ on the circumstances. In early summer of 1796, Napoleon led his troops into the city of Pavia during his victorious Italian military campaign. The author Pancaldi states that Volta was among the delegation of Pavia's prominent civic leaders who went to see the general to request that he protect their city. Napoleon, however, did not heed this request and the French soldiers pillaged Pavia.

An 1899 issue of *The Electrical World and Engineer* contains a story mentioning that Napoleon ordered his troops to spare Volta's house and even guard it from harm. Whatever the truth about Napoleon and his soldiers' activities in Pavia, the fact remains that Volta's encounter with Napoleon in 1796 proved to be beneficial for him in the years to come.

Near the end of 1796, Napoleon established the "client state," meaning a state that is under the rule of a more powerful country, of the Cispadane Republic (later called the Cisalpine Republic) in northern Italy, which included Lombardy. This was the government to which Galvani refused to pledge allegiance, resulting in his loss of professorship for a period. Volta was appointed to this short-lived government's legislature; however, he eventually resigned from this position out of loyalty to the Austrians.

Volta had a long history of working with the Austrian government. This country's leaders had financed Volta's trip around Europe, and the Austrians, in fact, did return to power in Lombardy in 1799 after France lost the War of the Second Coalition to them. As rulers, the Austrians were not an improvement over the French, as they closed the university and suspended Volta's pay; however, he was allowed to remain free. Rule of the Lombard region then switched back to French control in 1801 following France's military victories over Austria. The French subsequently re-opened the University of Pavia and Volta was given back his professorship.

This back-and-forth control of Lombardy between Austria and France undoubtedly would have made Volta quite leery about casting his lot behind either one of those nations. For instance, when Volta was considering where to make the announcement about his battery in 1800, Austria was the ruling government of Lombardy, but the following year it was under France's control.

Italy, obviously, was caught in the middle of the battle between the two nations—and was a prize that each country wanted to rule. Italy, indeed, was not a nation at that time, but more of a group of separate city-states.

The first Kingdom of Italy was created by Napoleon in 1805, and Italy wasn't unified until the mid- to late 1800s. Besides not being one of Europe's powers, Italy also did not have highly regarded science reputation. Consequently, Volta's homeland did not rank as a prestigious location for the announcement of his invention.

The remaining candidate was England. In the negative column was the fact that it was still in a long-running war with France. In England's favor, however, was that it inspired Volta to become something of an Anglophile. He had experienced many positive reactions and warm relations with British electricians and natural philosophers over the years—much more than he had with their French counterparts. In 1794, for example, he had received the prestigious Copley Medal from the Royal Society. That London also enjoyed the reputation as a major intellectual capital in Europe added another positive point for England.

## VOLTA ANNOUNCES HIS INVENTION

England won as Volta's final choice for the country in which he would break the news of his Voltaic pile invention. In the spring of 1800, he sent two letters to the president of the Royal Society, Joseph Banks. At that time near the beginning of his forty-plus year stint as president, Banks had been the Society's secretary when it presented Volta with the Copley Medal. In his correspondence with Banks, Volta described his invention as an apparatus that "will no doubt, astonish you." While this might have sounded like hyperbolic boasting, Volta's prediction was accurate.

It is interesting to note that Volta wrote his letters in French. It was a language that he knew very well, and it also served as a *lingua franca* in the science community. On the other hand, Volta used an English title, "On the Electricity excited by the mere Contact of conducting Substances of different kinds," in his first letter to Banks. These choices have been interpreted as an attempt to hedge his bets and appease his allies in both countries. In fact, Volta's political allegiances were worthy of speculation, because Britain and France were adversaries, and he had lived in an area under French control from 1796 to 1799. Considering his choices, Volta could be viewed as wanting to demonstrate respect to both France and England.

His first letter, dated March 20, was sent by mail and arrived in London around April 15. Instead of using the mail again, Volta gave his second letter, dated April 1, to a Como merchant named Pasquale Garovaglio, who brought it to London sometime before June 3.

This two-part letter approach was a deliberate strategy by Volta who was acutely aware of the political situation. His letters had to travel through France to get to England, and those two countries still were at war with each other. So, by sending two letters—by two different means—he hoped to increase the odds of their arrival.

The first letter was a mere four pages, and contained a 1,000-word description of the Voltaic pile. Choosing to avoid discussion of his invention's effects or include diagrams, Volta instead simply described the device and hinted at what it could do. This letter, consequently, can be seen as serving as an introduction—a way to alert Banks, and whet his interest. The second letter ran 5,000 words and included a sophisticated diagram of his invention, and discussed both the pile and the Crown of Cups device.

In his correspondence with Banks, Volta addressed four issues. First, there was the Voltaic pile itself. Second, the letters described how the battery was operated. The third point detailed how the device related to the controversy regarding galvanism. Lastly, Volta included an explanation of why the battery worked.

Addressing these points separately was intentional. During his nearly forty years as a scientist, Volta had unveiled other inventions—the electrophorus and the condensatore, for instance—and observed his peers' reactions. From these experiences, he understood how to maximize a positive reception by his colleagues. Typically, his fellow expert electricians had few issues in accepting and adopting Volta's electrical instruments; they were far more contentious when he sought to convince them of his explanations behind the construction of the instruments and how they worked.

In writing about the Voltaic pile performance, Volta chose to compare it to the well-known Leyden jars, but he wisely emphasized that what was different with his invention was that the pile did not have to be continually charged to generate electricity as did the Leyden jars. Furthermore, he used terms such as "perpetual impulse" to suggest the battery allowed electric flow of uninterrupted motion. Volta also underscored the instrument's simplicity of use, touting how the pile's discs could be made from materials that were readily available.

Volta additionally presented his new instrument as coming in two different designs: the pile (or column) battery as well as the Crown of Cups. While he regarded the two as equivalent, Volta noted that each one had strengths and limitations. Piles, for instance, were easier to construct and they were portable. The Crown of Cups battery, on the other hand, was more impressive in demonstrations since it operated in a visually interesting manner.

While Volta suggested that his invention came in additional

formats, it wasn't a point that he stressed. Instead he concentrated on aspects of the pile—the instrument and its performance—that were more easily understood. By mentioning the possibility of additional formats, however, Volta demonstrated that he viewed his battery as being part of "a family of instruments"—not just a single device—even in its early days.

He did not dwell on the theories behind this apparatus (nor even offered much of an explanation as to how the whole apparatus worked), but Volta did mention topics such as the torpedo fish's electric organ and the battery's effects on sense organs. These inclusions seemingly served as attempts by Volta to nip in the bud any controversy about whether or not the battery was his own invention. During this era, disputes about who had the original idea for an invention were common; Volta had been involved in some of them. So, he wanted to assert his ownership of the electric battery.

The reading of Volta's letter to the Royal Society occurred on June 26, 1800; however, it was not until later in the year that Volta's battery announcement actually appeared in The Royal Society's publication of record, *Philosophical Transactions*. Also, the journal decided to aggregate the two letters into a single paper for publication.

News of Volta's groundbreaking invention traveled very quickly throughout London. In fact, Londoners already knew about the Voltaic pile before either the Royal Society's reading or the publication of the *Philosophical Transactions* essay. By the spring of 1800 (late April and early May), several London experimenters had created and developed their own voltaic batteries.

An initial instigator behind this spread of information about the Voltaic pile was Joseph Banks, who told several acquaintances about Volta's invention shortly after receiving Volta's initial correspondence. One person he told was Andrew Carlisle, an esteemed physician at the Westminster Hospital. Carlisle, in turn, shared the news with his friend, William Nicholson. Yes, the same William Nicholson whose essay on the torpedo fish served as a major impetus for Volta.

Nicholson, in fact, was one of the only two or three fellow scientists to whom Volta referred in his letters. Pancaldi notes, however, that Volta's mention of Nicholson appeared in a separate part of the manuscript, and that Volta never really acknowledged that Nicholson's article was a factor in his creation of the pile.

Not surprisingly, Nicholson was very interested in Volta's invention. Partnering with Carlisle, the two men built their own Voltaic pile and started doing some experiments with it by the end of April. During their work, they observed chemical phenomena that suggested, to their thinking, water decomposition—a topic that Volta hadn't broached in his letters. Nicholson, however, did not publish these findings. He possibly made this decision because he was in a sensitive situation of having heard about Volta's invention before the Royal Society read Volta's letters at the end of June.

Nicholson, however, apparently did not refrain from talking about his experiments. The London newspapers devoted head-lines to this new electrical apparatus starting mid-spring of 1880. One report wrote about a May 28 presentation by Thomas Garrett, a lecturer at the recently established Royal Institution. His demonstration, which included a Voltaic pile as well as involving the decomposition of water, was deemed a success. The newspaper article, however, was so riddled with mistakes that it spurred Nicholson to pen a letter to the editor of a competing newspaper to set the record straight.

## WORD SPREADS ACROSS EUROPE

News about the battery continued to travel fast—particularly for 1800—faster than Volta undoubtedly anticipated. The battery took on a life of its own. Volta's latest invention was greeted with a great deal of excitement—some saw it as a transformational creation, and others rushed to replicate it. Pancaldi accurately characterized the battery as "a device that was almost instantly recognized as bringing about a momentous breakthrough."

Volta's invention became known throughout England, and news of it rolled quickly through many European cities. For his part, Volta instigated a personal publicity campaign in Italy. His friend and colleague, chemist Valentino Brugnatelli demonstrated a working Voltaic battery at a gathering held in his Milan home in April 1800. That same month, Volta contacted his go-to person in the court of Vienna, Marsilio Landriani, who also was one of Italy's leading natural scientists of that time. Excited about Volta's invention, Landriani, along with his Austrian colleagues, constructed their own version of a Voltaic apparatus soon thereafter.

As details about Volta's battery found their way through Vienna and all of Austria in May, people in Bristol (where Volta visited during his 1782 trip to Great Britain) and Glasgow, Scotland, learned of his invention in June. Word reached Haarlem in the Netherlands via Banks, who wrote of Volta's invention in a letter dated June 14, 1800 to the Dutch scientist Martinus van Marum (the man who had created the world's largest electrostatic machine in the 1780s). Copenhagen, meanwhile, became aware of the Voltaic pile later in that summer.

In 1800–1801, the Geneva-based *Bibliothèque Britannique, Sciences et Arts* printed several essays from Nicholson's journal.

These articles were printed in French, but they would later be translated into German by *Annalen der Physik*, a journal out of Halle in the Saxony region.

News of the battery started to reach France around the middle of August of 1800, when the *Moniteur universel*, the self-styled "official journal" of France, published an article on August 17 that reported on Garnett's London lecture. That *Moniteur* published the article was significant because the newspaper represented Napoleon's viewpoint. Its publication suggested that Volta was in favor with the Napoleonic regime. It was through the *Moniteur*, in fact, that Volta himself learned how the British were receiving his invention. The timing happened to be fortuitous for Volta because by the summer of 1800, the French once again controlled the Lombard region.

The popular press in both Paris and London played an important role in shaping the narrative of the battery's early days. The experts, meanwhile, shared their perspectives with each other through correspondence, personal contacts, and scientific journals. Just as the Londoners had done, Parisians (amateur and expert electricians alike) embraced the battery and conducted experiments within weeks after learning about it.

One Parisian who was greatly intrigued early on by Volta's invention was a man named Étienne Gaspard Robertson. A highly popular theater personality, Robertson staged a nightly live show entitled "Fantasmagorie de Robertson" that utilized a good deal of technology-based entertainment, such as employing magic lanterns to make ghosts appear, and one of the largest cranked friction machines in Paris to use in acts that featured electricity. On August 30, Robertson presented a report to the Institut National de France regarding his electrical experiments on the effects of voltaic batteries on the human body. His report later appeared in the *Annales de chimie*, making it the first article on the Voltaic pile to be published in a French scientific journal.

## VOLTA RETURNS TO PARIS

The spring of 1800 should have been a jubilant time for Volta as he had just created a device that would change the course of history. However, he could not celebrate his huge career triumph.

The Austrians, who were then in control of Lombardy, had not only closed the University of Pavia, but suspended Volta's university wages, which caused professional and financial hardships. These very struggles, however, would lead him in a profitable direction—to France, where he would achieve some of his greatest career triumphs. He did this, it should be added, without antagonizing the British—no small feat in a time of war and nationalistic fervor.

When the French regained control of Lombardy in the summer of 1800, Volta's situation greatly improved, since they reopened the University of Pavia and reinstituted his wages. The political instability in Lombardy made Volta uneasy, and he certainly couldn't feel confident about his future as a professor. Strategizing for his own career recognition and job security, Volta concluded that despite the political upheaval, the time was right to travel to Paris to campaign for his invention and for himself. Another factor indicating this was an opportune time for Volta was that he had just received favorable coverage in the *Moniteur*.

To help make this trip a reality, Volta wrote to General Guillaume Brune, the commander of the French army in Italy on September 28, 1800. Volta and the general had already been introduced, and Volta—ever the skilled politician—had talked to him about his invention. In his letter, Volta had two requests for General Brune. One was a leave of absence from his teaching position so he could travel to Paris and formally thank Napoleon for the University of Pavia's restoration; and second, that his family and friends would be protected while he was away.

Life in Europe, however, was still unsettled from ongoing political and military battles, causing Volta to remain wary from late 1800 and into early 1801 over the advisability of the Parisian trip. He did not want to find himself yet again on the wrong side of the political fence. His mind changed following the signing of the Treaty of Lunéville on February 9, 1801, which ended the war between the French and the Austrians. France remained at war only with Britain, although peace negotiations began in 1801 and concluded with the Treaty of Amiens in March 1802.

In the spring of 1801, Volta contacted Claude-Louis Berthollet, an important French chemist who would become vice president of the French Senate in 1804, and Gaspard Monge, a noted French mathematician who held positions in Napoleon's government. Both men had met Volta in 1797 in Como when they attended experiments he did on galvanism. To further lay the groundwork for his trip, Volta also sent a description of his battery to the well-known French chemist and government minister Jean Antoine Chaptal and to the distinguished geologist Déodat de Dolomieu.

Volta then requested a large sum of money for travel expenses from Milan's pro-French administrators. Realizing the political value for them in Volta's trip to Paris, the administrators granted his request for funds and facilities. On September 1, 1801, Volta departed for Paris with his colleague, the chemist Luigi Brugnatelli, who would later use Volta's battery to invent electroplating in 1805. Their first stop was Geneva, where Volta met with friends and began his campaign for the battery.

By late September they arrived in Paris, and Volta wound up remaining there for two months that autumn. He had two objectives for his visit to the City of Light—to gain some financial rewards and enhance his professional status.

To achieve the first goal, he sought to establish good relations with Napoleon, and to accomplish this he cultivated Minister Chaptal and Berthollet as two primary conduits to the French leader.

For his second goal, he wanted to demonstrate that he, and not Galvani, had been right in their great scientific debates of the 1790s, along with achieving an unblemished reputation among the Parisian scientific elite. By the end of his trip, Volta was successful on the first goal, but only moderately so on the second.

Still, this 1801 trip to Paris was an overall triumph, as Volta left with more recognition for his work and himself. Biographer Dibner eloquently describes Volta's visit as being "the beginning of a wave of rewards and praise that swept a modest and devoted professor into a position of honor reserved for the immortals of all time, and rarely given to a man of science."

Volta started his mission as soon as he arrived in Paris. In his very first days in the city, Volta met with Berthollet, Robertson, and several other important members of the Parisian science community and society. His days in Paris were filled with performing experiments, meeting with dignitaries, taking meetings, attending soirées, and making new acquaintances, as well as renewing old ones. On October 2, Volta and Brugnatelli were honored by being named as members of an Institut Nationale commission on galvanism. Throughout it all, the

Lombard professor campaigned vigorously, and with a good deal of success, for his battery.

## VOLTA MEETS NAPOLEON

As part of his strategy to make contact with the powerful Napoleon Bonaparte, Volta dined with General Brune at the Ministry of War on October 6. The following day, he met Minister Chaptal, and then ten days later, he had lunch with Chaptal again. Over the course of his first month in Paris, Volta networked so successfully that by November 6 he received an introduction to Napoleon.

The following day, November 7, marked a major turning point in Volta's career and his life. At the Institut National de France that day, Volta delivered his first lecture about the battery, and Napoleon was in attendance for an hour of it.

During the month of November, Volta would present two additional lectures at the Institut National. These talks, on November 12 and November 22, served to showcase his latest invention and afforded Volta the opportunity to reintroduce himself to the Parisian elite, following the less-than-enthusiastic reception he had received in 1782. Helping to assure that he would receive a more positive reaction this time was the fact that the imposing First Consul of France, Napoleon Bonaparte, attended all three lectures.

Besides revealing the features of his Voltaic pile, Volta also performed several experiments and demonstrations. As a public scientist and showman, he captured the audiences' attention by exploding his electric pistol, and generating sparks and shocks with a large battery composed of eighty-eight pairs of metallic discs. He also smartly involved Napoleon as a participant in his presentations, which included melting a steel wire, decomposing water, and creating electrical sparks. Volta put on compelling performances for the First Consul and the whole audience showered him with applause.

In a letter to his older brother, Volta wrote about Napoleon and his time in Paris:

> *Bonaparte was in good humor, at ease and gracious, and the conversation lasted more than an hour and a half. I, myself, joking aside, am amazed how my old and new discoveries of the so-called galvanism, which show them to be only pure and simple electricity caused by the contact of metals, could have produced so much excitement. Objectively regarded, I find them also of some importance, they certainly throw new light on the theory of electricity. They open a new field for chemical research and also offer applications to medicine. For a year or more the journals of Germany, France and England have been full of discussions about them. Here in Paris there is the current excitement where, as in other things, there is added the excitement of fashion.*

Napoleon was won over by Volta from his very first lecture. When it concluded, Napoleon turned to his physician, Corvisart, and exclaimed enthusiastically, "Here, my good doctor, we have the image of life itself! The pile represents the column of vertebrae, the liver is the negative pole, the kidneys, the positive pole!" Putting his words into action, Napoleon proposed that the Institut National honor Volta with a gold medal. Because he was a member of the Institut, as well as the powerful leader of France, Napoleon understandably held considerable influence in making such a request. Even so, the other Institut members demonstrated some initial reluctance to agree with Napoleon's request.

Volta's relationship with the Parisian scientific elite had been an uneasy one since 1782. He didn't improve his standing in 1801 either when he failed to mention the work of the celebrated French physicist Charles-Augustin de Coulomb while citing his own condensatore and straw electrometer during his battery demonstrations. Volta further stoked the indignation of the Parisian science community by ignoring Coulomb's principles of torsion balance, which could have put Volta's invention on more solid scientific footing.

Ultimately, though, Napoleon's enthusiasm carried the day. He wanted Volta to be so honored, and the Institut National, albeit with some reservations, fell in line with their leader. Volta received his gold medal.

Napoleon placed a high value on science, and did so for a variety of reasons. In their biography of Napoleon, the famous historians Will and Ariel Durant hailed him as "the first modern ruler with a scientific education [who] restored Louis XIV's practice of awarding substantial prizes for cultural achievements, and he revealed his background by giving most awards to scientists." They also noted that Napoleon did not just honor French scientists but foreigners too, and specifically mentioned Volta as one of the foreign scientists that he honored.

An intellectually engaged and curious man, Napoleon appeared to have held genuine respect and admiration for Volta; however, the future French emperor also honored him for pragmatic motives. One reason behind his support of

Volta was that Napoleon thought that there might be some technological—and even military—uses for the battery that could further his ambitions.

Part of Napoleon's political agenda was to elevate the French image in Europe, and one way he saw to do this was to promote Paris as the intellectual center of Europe for all nationalities and political persuasions. By 1801, Volta had acquired name recognition among people who mattered to Napoleon—the "cultured elites of most European capitals."

While Napoleon had multiple reasons for wanting to recognize Volta, Volta had reasons for wanting that recognition. In saluting Volta, the Napoleonic regime could align itself with the movement of scientific progress in vogue at the time, which could counter the skeptical or downright hostile attitude that many Europeans then held for the French government. Napoleon derived both political and diplomatic benefits from his celebration of Volta. Even though Napoleon's actions were partially a publicity maneuver, the Lombard scientist was happy to participate, so long as doing so was in *his* interests too.

The cordial relationship between Napoleon Bonaparte and Alessandro Volta was mutually beneficial. Napoleon wielded hard power, as some would describe it today, while Volta symbolized soft power. Napoleon wanted to buttress his hard power with the soft power of cultural influence held by Volta. It was not, of course, a relationship of equals; Napoleon represented the patron and Volta was just a successful scientist, but he served happily as a mechanism of the French ruler's desire to enhance his soft power.

By the autumn of 1801, when Volta took his journey to France, peace had arrived in Europe, and on favorable terms to the French due their military victories. This turn of events allowed Volta to ally himself with Napoleon without the risk of alienating his friends in Britain or any other parts of Europe. Volta's Institut National appearances, moreover, coincided with several major Parisian events celebrating the recent peace among the European powers. It was during his stay in Paris that Volta was also chosen to represent the Cisalpine Republic at the upcoming congress of the Italian Constitutional Assembly, a large, influential group over which Napoleon wanted to maintain his power.

After more than two months in the City of Light, Volta left on December 4 to travel to Lyon. Positive news coverage of Volta continued in Paris even after his departure. Another sign that his trip to Paris was a success materialized just days after his departure, when the Consuls of the French Republic awarded Volta with a gratuity of 6,000 francs.

Napoleon was behind this decree and six months later he honored Volta again in announcing the following:

> *As encouragement to further experimentation and discovery, I wish to give the sum of 60,000 francs to the one who will give to electricity and galvanism the advances in this field equivalent to those already given to these sciences by Franklin and Volta. My special aim is to encourage and fix the attention of physicists on this branch of physics which is, in my opinion, the road to great discoveries.*

Despite these honors and Napoleon's embrace of him, Volta still had to work to win over the Parisian natural philosophers in 1801. Without the backing of the powerful Napoleon, the Institut National probably would have been far less generous in its honors and recognition. Indeed, Volta never fully won over the Parisian scientific elite, and he had little contact with them after his triumphant 1801 trip. Even though he did not totally achieve the enthusiastic support of the Institut National, Volta left Paris having successfully gained more financial rewards and a more prominent professional status.

## A LIFE OF ITS OWN

To achieve popular acclaim for his Voltaic pile, Volta had to do more than convince the small group of expert electricians of its importance. He needed to impress a large and diverse group of people who lived different countries and came from a variety of cultural and social backgrounds. Volta was able to accomplish this through his own self-styled "publicity campaign" of 1800–1801, which brought consequences that he probably could not have foreseen, and definitely could not control, whether he wanted to or not.

News of this electric battery sparked immediate interest in people across all levels of social, economic, and professional groups. Consider, for a moment, the trickle-down effect of Volta's first letter about the Voltaic pile. It was sent to the esteemed head of the Royal Society, Joseph Banks. The man to whom Banks passed Volta's letter was not a scientist but a physician, Anthony Carlisle. While Carlisle did tell the scientist William Nicholson about the battery, Nicholson was not part of the elite British scientists who belonged to the Royal Society.

One reason the Voltaic battery generated such instantaneous interest was that it was a relatively simple device. It could be constructed by an electrician with a little skill and knowledge. The materials needed to build one were not hard to find, and the battery's design made it simple to construct. Enterprising professional scientists, as well as curious amateurs, set off in a mad rush to amplify and transform this invention. The battery, consequently, invited many different uses and interpretations. Eventually, there would be batteries in bigger sizes and varying formats. Volta's invention, coupled with chemical advances, evolved into devices that were far more complex and sophisticated.

Because the battery was so new, unique, and groundbreaking, experts and amateurs disagreed about its meaning, functionality, and guiding philosophy, but they all appreciated and acknowledged Volta's remarkable history-shaping invention and its role in the creating "The Battery Story."

The battery, in other words, took on and developed its own identity—divorced from what its inventor may have intended. Going back to the electrophorus and his subsequent inventions, Volta had seen how scientists and laypeople alike would formulate their own conceptions for these devices, some of which had little to do with the interpretations of natural philosophers.

Volta is said to have been somewhat disappointed that the battery so quickly took off in a direction different from what he had foreseen—and he could no longer direct the narrative of his own invention. Controversies, like the one prompted by Nicholson concerning Volta's failure to address the battery's chemical phenomena, undoubtedly concerned him.

Conversely, Volta also knew from the electrophorus's complicated reception in the 1770s that his theorizing about the battery would lead to difficult discussions, debates, and negotiations among his peers. This was a reason why he decided to focus on what the Voltaic pile could do, instead of offering explanatory theories.

For Volta, it was more important to avoid major controversies surrounding his claim of being the inventor of this battery, a problem that had arisen with some of his earlier inventions. That others appropriated the battery for their own uses did not significantly upset Volta because he was still acknowledged as its originator. It would have been an ill-advised move on Volta's part to insist too strongly on controlling what could or could not be done with a battery, since he would risk undercutting all the popular positive sentiment his invention had engendered. It could easily be said that he was happy just to bask in the radiant glow of the accolades and material rewards that came his way, and to stay out of the intellectual debates and philosophical discussions of the battery.

Everyone, it seemed, had an opinion about Volta's attention-grabbing electrical device, and reactions overwhelmingly were positive. Much eager anticipation—and very little trepidation—greeted the battery's arrival in Europe. Unlike some inventions, Volta's battery didn't really generate significant opposition toward its use or role in society. That so many people were talking about the battery was a sign of its transformational power. The vast majority clearly viewed it as a sign of progress, and one to be embraced.

The battery's narrative was shaped by the public as much as by the European community of inventors and scientific enthusiasts. Volta received credit and celebrity because of his invention, but he eventually became rather incidental to its future, as other experts and amateurs built on and improved his original Voltaic pile.

## WHAT'S IN A NAME?

Every invention needs a name, one that helps it to achieve widespread acceptance and recognition. The names that ultimately become the most popular are those with the broadest appeal, cause the least debate and are not connected to any specific group or interpretation. This was very much the case in regard to the word "battery" as the accepted name for Volta's creation.

Volta's experiences with the electrophorus a quarter century earlier proved to be a contributing factor in the naming too. Knowing full well how the right name could play a crucial role in helping others to track and recognize a device's value, Volta paid close attention to what his invention would be called. "New instruments should be given new names, depending not only on their form, but also on their effects, or the principle on which they are based," he told the Royal Society in 1800.

The first two potential names that Volta came up with for his signature invention were *organe électrique artificial,* or artificial electric organ, and *appareil électro-moteur,* or electro-motor. Neither one, however, was adopted in any appreciable way. The device that had quickly become the property of the world at large, needed a broadly understood, not overly complicated name, and the eventual word for Volta's device was entirely divorced from him and his perspective.

The term *battery* came to the fore—and soon achieved linguistic immortality—for several reasons. There had already been an established connection between the word and electricity, going back more than half a century. Expert and amateur electricians alike were familiar with "the batteries of Leyden jars" dating to Franklin's research during the 1740s. This association also appeared in all sorts of literature, from manuals to treatises.

Inspired by the usage of the word "battery" in military context, referring to "a set of similar pieces used for combined action," electrical experimenters began referring to Volta's instrument as a "battery" because it similarly consisted "of many identical metallic pairs combined together." As early as 1803, British textbooks about electricity included entries about "Volta's batteries," "galvanic batteries," and "voltaic batteries."

While Volta invented the battery, and the Voltaic pile carried his name, he curiously showed no interest in patenting it, even though the patent system already existed in European countries. At the beginning of the 1800s, a person had to show the practical applications and the expected profits to get a patent for an invention, and apparently Volta did not see his device fitting into either category. In those days, moreover, the patent process did not offer a surefire path to victory for Volta. Instead, he pursued the more certain path of recognition and renown as the way to reap rewards.

Volta's strategy worked—he did achieve continent-wide celebrity and the accompanying benefits. He must have felt a definite satisfaction in those post-Paris days of 1801 and 1802. The drive and ambition that Volta exhibited as early as his teen years—when he felt a bright future awaited him—assuredly paid off four decades later.

Volta certainly deserved every honor, reward, and accolade that resulted from his invention of the battery that has, and continues to have, a far-reaching effect on the world.

# Chapter Ten

## VOLTA STARTS A REVOLUTION

### A GAME CHANGER

"Modern science was founded by [William] Gilbert, who discovered the magnetism of the earth. It was vastly extended by Franklin, who discovered the electricity of the atmosphere. Volta furnished the world with a new source of the electric fire." These are the words of John Munro in the preface of *Pioneers of Electricity*, which he wrote in 1890. While this was written nearly a century after the initial excitement of the invention of the Voltaic pile, it shows that the magnitude of Volta's creation was still was recognized.

Volta's battery occupies a rarefied place in the modern history of science and technology, and Volta still is acknowledged for this milestone invention. The historian John L. Heilbron has described Volta's crowning glory as a discovery that "opened up a limitless field" and "transformed our civilization," while praising the pile as "an epoch-making prime mover." Because it represented the first time a continuous supply of electricity had become available from a simple source, the Voltaic pile continues to be roundly hailed as central to revolutionizing nineteenth-century science and technology. *Scientific American* editor Scott Fletcher proclaimed in his book *Bottled Lightning* that "Volta earned such effusive praise because of the battery's enduring, history-bending influence."

## THE BATTERY AND THE INDUSTRIAL AGE

Many advances in machines and devices became everyday conveniences that humans now take for granted were first developed during the decade surrounding Volta's invention of the battery.

The list below includes a selection of inventions that occurred in the period 1795–1805.

## MAJOR INVENTIONS BETWEEN 1795–1805

- 1795: Hydraulic Press
- 1796: Vaccination
- 1797: Cast Iron Plow
- 1797: Screw-Cutting Lathe
- 1798: Lithography
- 1799: High-Pressure Steam Engine
- 1799: Battery
- 1800: Nitrous Oxide Anesthetic
- 1800: Submarine
- 1802: Gas Stove
- 1804: Locomotive
- 1804: Municipal Water Treatment

- 1804: Glider
- 1805: Endoscope
- 1805: Electroplating

Source: Adapted from data in Jack Challoner, ed., *1001 Inventions That Changed the World* (Hauppauge, N.Y.: Barron's Educational Series, 2009, 236–57.)

That Volta's battery arrived just as the Industrial Age was taking off at the end of the eighteenth century is no coincidence. The invention played a fundamental role in the rapid scientific and technological progress that occurred during this transformational period.

During the nineteenth century, scientists as well as the public were enthralled with the concepts of energy. Jürgen Osterhammel, in his landmark book, *The Transformation of the World: A Global History of the Nineteenth Century*, credited Volta's experiments as leading "to a whole new science of energy, and various cosmological systems" by the middle of the century.

The pivotal scientific work involving the battery took off within weeks and months of people learning about the Voltaic pile, with experimenters all across Europe copying Volta's inventions and creating batteries that were bigger and stronger.

Nicholson and Carlisle were among the first to experiment with batteries. They each constructed a battery pile featuring thirty-six half-crown silver coins and zinc discs in 1800, just weeks after Banks received Volta's letter.

Many others soon followed suit. In the English town of Bristol, Thomas Beddoes added a Voltaic pile that consisted of 110 pairs of metal plates to his famous Pneumatic Institute, while Scottish physicist John Robison put together a battery featuring seventy-two pairs of metal discs.

During the summer of 1800, the Parisian showman and amateur physicist Étienne Robertson built a *colonne métallique*. Because of his great interest in galvanism, Robertson even tested his machine on body parts to see what the reactions would be.

In Germany, Johann Wilhelm Ritter began exploring the powers of the battery in the fall of 1800. The twenty-three-year-old Ritter, whose interests included both science and the Romantic philosophy of nature, had begun corresponding with Volta two years earlier about galvanism. His early work included experiments on the human body like Robertson's and he delved into chemical matters just as Nicholson and Carlisle had done.

One of the first significant advances in the battery occurred in 1802 when the Scottish chemist William Cruickshank improved Volta's battery design by making it horizontal instead of vertical. Cruickshank placed square sheets of copper and zinc horizontally in a rectangular wooden trough, or box, that was filled with an electrolyte such as brine and sealed with shellac. The advantage of his "trough battery" was that it eliminated most of the electrolyte leakage that was prone to happen with a vertical pile construction. Cruickshank's battery also was the first that could be mass produced.

At this time, the British chemist-inventor Humphry Davy was so inspired by Volta and the potential of the Voltaic pile that he assembled a battery composed of two thousand cells which has been described as the largest, most powerful battery of its day. Located in the basement of London's Royal Institution, this giant battery was used as vital part of Davy's pioneering research into what is now known as electrolysis.

By splitting common compounds, Davy was able to discover such new metals and elements as magnesium, sodium, and potassium. He also used the large Voltaic pile when he invented the first practical electric light. Quite popular from the mid-nineteenth century and into the twentieth century, Davy's creation opened the door for many technological advances for lighting. While typically known as the carbon arc light, it was also known as the Voltaic arc in recognition of Volta's influence in its creation.

Davy also was not shy about acknowledging Volta's importance. In his first Bakerian Lecture at the Royal Society in 1806, he opened with the following remarks:

*It will be seen that Volta has presented to us a key which promises to lay open some of the mysterious recesses of nature. Till this discovery, our means were limited; the field of pneumatic research had been exhausted, and little remained for the experimentalist except minute and laborious processes. There is now before us a boundless prospect of novelty in science; a country unexplored, but noble and fertile in aspect; a land of promise in philosophy.*

In 1814, Davy even traveled to Italy to see the elderly Volta and pay his respects to the science pioneer. Joining him on this trip was his assistant Michael Faraday, who was presented a battery from Volta during this visit. Faraday would later become famous for his work with electricity and magnetism. It was Faraday who discovered how circular movement between a magnet and a coil can create endless electricity. Faraday's principle of "electromagnetic rotation" paved the way for such pivotal inventions as the electric generator and the electric motor. His discoveries in electrochemistry, known as Faraday's

Law of Induction, also led to the ability to construct large transformers and other vital machinery that helped to power the Industrial Revolution of the nineteenth century. Faraday's research has been described as having much in common with Volta's, and Faraday himself praised the Voltaic battery as being a "magnificent instrument of philosophic research."

Another famous scientist influenced by Volta's work was the Danish physicist Hans Christian Oersted. After learning about the Voltaic pile, Oersted not only wrote about it, but started experimenting with one as well. His major discovery, that electric current also creates a magnetic field, was made in 1820. It was a connection that had not been made before, which Debner found "a mystery [as to] why this phenomenon evaded the experimenters on both sides of the Atlantic for so long after Volta's discovery."

Oersted's revelation, however, didn't remain unknown for long. The French mathematician-physicist André-Marie Ampère (who nearly twenty years earlier had been greatly inspired by a Volta speech he heard) learned about Oerstad's discovery and developed it into a theory of electromagnetic motion that he called "electrodynamics," but which is now better known as electromagnetics.

Electromagnetics are presently vital in the construction of machinery such as transformers, MRI scanners, and metal detectors. In the nineteenth century, however, its chief contribution was in helping the electric telegraphy come to life. Also, if Ampère's name sounds familiar that is because he was later memorialized by having the basic unit of electric current christened the "ampere," or "amp" as it is commonly used today.

The telegraph, as writer and historian Sean Trainor wrote in an article published on the *Time* magazine website, "represented a revolution in communications rivaling both the printing press and Internet." The invention of the telegraph marked the first time that messages did not have to be delivered physically, by human, horse, or bird. Communication, consequently, became far less constrained by distance. Consider how many weeks it had taken Volta's letter about the invention of the battery to reach Joseph Banks, or how many months it took Volta to receive the books he ordered that led him to that invention. The telegraph transformed society in its time.

Besides revolutionizing communications, the telegraph additionally made an impact on the world economically. By the mid-1800s, the growth of telegraphy altered the employment landscape, with young men leaving manual labor jobs like farm and mining for jobs that were more office-related. This shift also reflects the overall movement of people relocating to cities from rural areas during the nineteenth century.

Behind all this progress was the battery. As author and historian Adam Hart-Davis put it, "Only after the Italian scientist Alessandro Volta discovered a means of generating steady current in 1799 that an electric telegraph became a real possibility."

The 1800s were years in which many meaningful technological innovations and scientific inventions powered by electricity came into being. Interest in inventions extended beyond the scientific and academic communities to the general public. Exhibits showcasing new inventions and the latest technology were quite popular throughout this century. It is no coincidence that the first World Expo took place in 1844 with the French Industrial Exposition. Seven more took place by 1899 and, in addition, three more were held between 1900–1904.

The nineteenth century, in fact, was filled with a huge number of technological "firsts": the first transformer (1836); the first microphone (1860); the first transatlantic cable (1866); and the first thermal power stations (1882).

The list below spotlights some of the advancements that occurred simply between the years 1876–1880.

- 1876: Electric Carbon Arc Lamp
- 1876: Telephone
- 1877: First Street Lighting Installed in a City (Paris, France)
- 1877: Phonograph
- 1877: Carbon-Button Microphone
- 1878: First Hydroelectric House (Cragside, England)
- 1878: Incandescent Lightbulb
- 1879: Electric Rail-Train
- 1879: Long-Lasting Filament for the Incandescent Lamp
- 1880: Piezoelectricity Discovered

Sources: Britannica.com; americanhistory.si.edu; famousscientists.org; dailymail.co.uk; and wired.com.

This era experienced such a boom in the uses for the electric battery that by 1878 a French electrician named Alfred Niaudet published a book featuring several hundred different types of batteries.

From electrolysis to the telegraph to the electric tattoo machine, which debuted in New York City in 1891, Volta's invention held a wide-ranging influence on scientific advances in the nineteenth century. Chemists used it. Physicists used it. Inventors used it. Scientific generalists used it. The battery unquestionably acquired an importance that crossed national boundaries and disciplinary lines.

Although the long-time University of Pavia professor might not have seen any of his own students become notable figures in the science world, Volta served as an influence and inspiration to many fellow scientists across Europe—Nicholson, Davy, Ampère to name a few—who carried on, and expanded upon, his work. This multi-national connection between Volta and the next generation of scientists is fitting, considering how Volta worked so hard to meet and get to know scientists from European countries when he was a young man looking to make his mark on the world.

## WET VERSUS THE DRY BATTERY TYPES

The electric battery grew in significance to the science and technology worlds during the nineteenth century. As its uses continued to expand, the battery continued to evolve.

Batteries today are made one of two ways, as a wet-cell battery with a liquid solution or a dry-cell battery with a dry paste. They can also be classified as a "primary cell" (which works only as long as there's current stored in the battery) or a "secondary cell," which is rechargeable. Early batteries all were "wet cell" and "primary cells," but technological enhancements have led to the creation of the "dry cell" and the "secondary cell," which also have expanded the ways that batteries can be used.

Volta's invention represented the first "wet cell" battery, and the electrical experimenters who followed him also created batteries that fit into this category. One improvement was that they tended to use acid as the electrolyte—or liquid solution—

between the electrode (the two metals) in the battery rather than brine (the electrolyte used in the Voltaic pile). Many of these early batteries were made with glass jars, which made them vulnerable to breaking.

A battery's life greatly improved in 1836 when an English chemist and meteorologist named John Frederic Daniell created the next-generation wet-cell battery that was less corrosive and safer to use than the Voltaic pile model.

Daniell placed a zinc rod electrode, which had been dipped in diluted sulfuric acid, into a porous clay pot. He then put this pot into a copper container filled with a copper sulfate solution. The electrical current that flowed from the copper container (functioning as the positive electrode) to the zinc rod (serving as the negative electrode, or the cathode) was more stable than the current generated in the Voltaic pile. Additionally, Daniell's changes fixed the polarization problem with Volta's design, where hydrogen bubbles would build up on the copper (positive) electrode to the point of impeding the electron flow and consequently shutting down the battery.

While the "wet cell" provided many significant contributions during the first two-thirds of the nineteenth century, this battery's inherent shortcomings limited its uses and practical applications. The conductive solution typically was some variety of acid, which made it dangerous to touch. The battery's heavy weight made it difficult to move, and the electrodes frequently fell out if the battery moved too much. Moreover, the battery was prone to spilling when it fell over.

In 1866, Georges Leclanché introduced the next major leap in battery construction when he designed the first type of dry cell-style battery. The French engineer succeeded in reducing the battery's toxicity when he changed the electrolyte to an alkaline (ammonium chloride) instead of an acid. Furthermore, he was able to make a more lightweight battery by substituting the lighter carbon-manganese dioxide for copper as one of the electrodes. Because it was enclosed with a hard-wax mixture, this battery offered another advantage: it was less likely to leak than other versions.

The following year, the German scientist Carl Gassner conceived a truly "dry" cell battery. By creating a paste composed of plaster of Paris and liquid ammonium chloride, Gassner could eliminate the need for a liquid electrolyte. He also used a sealed metal case, which made his battery easier to move and carry as well as being far more convenient to use. The dry-cell battery enabled electrical devices to be portable, which made them indispensable items for modern living.

While their constructions are similar, primary- and secondary-cell batteries function in very different ways. Both types process materials in the electrolyte and the electrodes, so that chemical energy turns into electrical energy. But primary-cell batteries stop working once they are completely discharged; these batteries are commonly used in toys. Secondary cell batteries are rechargeable.

The way a rechargeable battery works is quite simple. Once a secondary cell fully loses its charge, the battery can be attached to a type of direct current and recharged. In 1859, the French physicist Gaston Planté was the first person to invent a practical rechargeable battery. His design utilized an acid and lead system, which is still common today. The popularity of rechargeable batteries increased with the advent of the nickel-cadmium battery that the Swedish scientist Waldmar Jungner invented in 1899. Nickel-based secondary-cell batteries offer longer life and more power, but they create environmental problems.

Today, secondary-cell batteries, also known as storage cell or accumulator batteries, power a wide range of items, from cell phones and laptop computers to vehicles. The rechargeable battery and other nineteenth-century advancements in battery technology set the stage for an explosion of uses during the twentieth century, as the battery came to power essential devices of our modern life.

## BATTERIES AND THE MODERN AGE

"It's hard to think of another technology as foundational as the battery for how we live today," asserts Harold Wallace, the associate curator of the Electricity Collections at the Smithsonian's National Museum of American History in Washington, D.C. A dynamic invention, the battery epitomizes an "enabling technology" that intersects with all kinds of important historical, scientific, and technological dynamics. Batteries are so much a part of our lives—from cellphones to space shuttles—that their absence would remake our very existence.

When *Smithsonian Insider* journalist Michelle Z. Donahue listed the five batteries that revolutionized the world, she placed Volta's pile as the first and oldest battery, with the other four listed all having been invented in the twentieth century. What are the other four revolutionary batteries on her list?

One is the Rechargeable Nickel-Iron Battery, which was devised at the turn of the century by Thomas Alva Edison. Yes, the man best known for inventing the phonograph, the light bulb, and the movie camera also built an electric battery for automobiles. Electric cars are typically associated with the twenty-first century, but they also achieved some popularity in the early twentieth century. Although Edison's batteries failed to transform the American car industry, his nickel-iron battery design has been used for years to power railroad signaling and forklifts because they have an extensive life span even when they have been idle for a long time.

Another historic battery on this list is the Mercury Button Cell Battery, which was invented in 1942 by Samuel Ruben. While its primary use was for military equipment such as walkie talkies and metal detectors, these compact batteries also proved very popular in the commercial world to power toys, hearing aids, and watches (the first electric watches, for example, were produced in the late 1950s). Originally, these batteries contained mercury as an electrolyte, although they are made now with less toxic elements such as lithium and zinc.

Like the Mercury Button Cell Battery, the Proximity Fuze Battery was created in 1942 for military purposes. The U.S. Armed Forces developed this battery because it needed a small, secure power source with time-delay capability to place into missiles and bombs. These batteries wound up playing a pivotal role in World War II and later in the Korean War effort.

The fourth important battery was created in the late 1950s by a Union Carbide chemical engineer named Lewis Urry, who improved upon Edison's use of zinc in an alkaline battery by substituting it with powdered zinc. With this new electrolyte, Urry devised an alkaline battery that had more longevity than any previous zinc-carbon battery. It was originally known as the Eveready Battery, and now is popularly known as the Energizer Battery.

Another essential battery that did not make the Smithsonian Insider list is the Lithium-Ion Battery, which was developed in the 1970s and 1980s. This type of battery is found in common household products such as power tools and game consoles. Extremely versatile, the Lithium-Ion Battery can be manufactured to be small enough to fit into smart phones, digital cameras, and computers or sized large enough for use in electric bikes, electric cars, and even the Mars Rover.

## THE FUTURE

One of the most famous names currently associated with batteries is the renowned entrepreneur Elon Musk. Lithium-Ion batteries, for example, serve as the main power source for his Tesla electric vehicles. Furthermore, Tesla is involved in the battery business. The company has public supercharger stations that charge electric cars and has developed the Powerwall, large Lithium-Ion battery packs for home and industrial use. Tesla's plans also include working on a large lithium-ion battery as part of the company's project in South Australia that is designed to be a "wind battery farm."

Wind, solar, and hydrogen have all been mentioned as possible sources for batteries in the future. In writing about how our society takes batteries for granted, Ireland-based science journalist Sean Duke recently asserted that unlike microchips, "battery technology has changed little since Alessandro Volta's stacked battery," even though its uses have expanded tremendously. Indeed, the dry-cell battery has remained much the same—an electrode surrounded by paste and sealed in a metal case—since the days of the Voltaic pile. While meant as a warning, Duke's comment can also be seen as a compliment to the genius construction and enduring nature of Volta's landmark invention.

Another example of the ongoing relevancy of Alessandro Volta's experiments is *The Atlantic* story, "A New Kind of Soft Battery, Inspired by the Electric Eel," by Ed Yong. Volta is mentioned in the opening sentence of the article, which examines researchers at the University of Fribourg in Switzerland who, much like Volta, were inspired by the physiology of the electrical eel for their design of a battery. With the advantage of more than two hundred years of advancements since Volta's days, these researchers have devised a model that more closely imitates the eels' electrifying organs. While Volta built his battery out of two dissimilar metals separated by a wet material, this Swiss team has utilized cutting edge technology to create a potentially revolutionary power source composed of "blobs of gels." This type of "soft battery," which is activated merely by pressing the gels together, could provide dynamic ramifications in the future for medical equipment such as pacemakers and prosthetics, and robot design.

# Chapter Eleven

## VOLTA'S LIFE AS A LEGEND

### THE MAN OF SCIENCE, AFTER 1802

Alessandro Volta reached a pinnacle of success during his 1801 trip to France. Not only were his Voltaic pile presentations enthusiastically received, but Napoleon Bonaparte showered Volta with accolades—and made sure he was elected into the French Institute, awarded him a gold medal, and arranged for him to receive 2,000 crowns for traveling expenses.

Volta returned home to Lombardy with the professional success and stature that he had worked for. Indeed, his life would never be the same again. The honors and recognition Napoleon conferred on him in 1801 helped to transform his reputation to nothing less than a scientific legend in Europe.

Had Volta achieved this level of fame today, he might have capitalized on it in ways celebrity scientists such as Neil deGrasse Tyson or Bill Nye have—by writing books, hosting TV shows, and traveling on lecture tours. The Lombard scientist, however, did just the opposite. Instead of building upon his newly achieved prominence, he turned his attention to teaching and academia, and to life with his family.

Even more surprising, Volta virtually stopped conducting research and experimentation. Before 1800, for example, he had issued more than eighty articles, letters, and books, either in their original language or in translation. After 1800, he published very few pieces, and none of them dealt with subjects of much import. Volta's meager output of scientific work has surprised scholars, even those writing in the nineteenth century. Munro, in his book *Pioneers of Electricity*, remarked how Volta "amused himself with trifling experiments in galvanism on dead animals, such as making a cricket sing, or increasing the lustre of a beetle; while Carlyle [nee Carlisle] and Nicholson were decomposing water, and Humphry Davy was separating the alkaline metals, sodium and potassium, from their earths."

Volta's retreat from science has generated a good deal of theorizing. Some believe that Volta was exhausted from the tremendous amount of work he did in the prior few decades. Others hypothesize that his shortcomings as a mathematician hampered him in the increasingly complex field of electrical research. An article in the 1903 volume of the journal *The Electrical Engineer: An Illustrated Record and Review of Electric Progress* commented that Volta "was not strong in mathematics, and when these became necessary for further progress he wisely stood aside." Volta himself seemingly acknowledged this point when he commented how numbers and degrees are necessary for something good to be accomplished in physics. The well-respected French physicist and mathematician Jean-Baptiste Biot, who chaired an august commission organized by Napoleon on electricity, noted that

Volta's lectures often avoided the mathematic side of physics, as well as optics. Similarly, his published works did not contain much math.

While Munro mentions Volta's limitations as a mathematician and theorist, he also praises his power of perseverance, his skills as an observant experimenter, and his gift of genius. These attributes are apparent in the number of Volta's original inventions as well as his talent for being able to examine an idea or invention and improve upon it—whether it was his own or someone else's, as in the case of reading about the torpedo fish in Nicholson's article and allowing it to serve as an impetus for creating the Voltaic pile.

Volta's backing away from science experimentation can also be viewed as a case of simply feeling that he had nothing left to prove. All the hard work, strategizing for success, and belief in himself had paid off. The driven, ambitious young man who had wanted to make his mark on the world had accomplished just that; he had achieved professional successes and his legacy was assured. "The battery assured Volta a place in the pantheon," writes Seth Fletcher in his book, *Bottled Lightning*.

While Volta appeared to have been content with his decision to focus his later years on his wife and children, his choice represented a significant shift in priorities. This was a man who, as a young man, yearned to leave the stifling confines of his hometown and then spent nearly forty years concentrating on his career with apparently little thought of having a family life.

## VOLTA'S WILD BACHELOR DAYS

Before getting married, Volta had a reputation for his rather active social life. Georg Christoph Lichtenberg, whom Volta visited during a trip to Göttingen in 1784, described his colleague from Lombardy as a man who "during some extremely uninhibited hours," would talk well past midnight and reveal "an expert knowledge of the electricity in girls."

In his letter, Lichtenberg also included a stick-figure representation of Volta that depicted him as being in what can be termed as a state of arousal. Lichtenberg's description and frankly off-color drawing of Volta suggests that he may not have been the work-only, straight-laced scientist as a young man.

During his bachelor days, Volta had significant relationships with two women. The first was Teresa Ciceri, a member of a prominent Como family. Five years younger than Volta, Ciceri has been described as "energetic and lively, with a series of cultural and practical interests," qualities similar to those possessed by the woman Volta would eventually marry later in life (and whose name was also Teresa). If he had been interested in marrying Ciceri, Volta missed his chance because she married at age twenty in 1770, some thirty years before he apparently was ready to marry. Volta, at the time of Ciceri's marriage, was far more focused on building his career, which was then still in its early stages. He had only published his first treatise on electricity the year before.

Whether or not Volta and Ciceri had a love affair is only speculation, but it is well-documented that the two did have a life-long, close friendship. It is worth noting that strong friendships between men and women were not unusual during this time period. For his good friend, Volta would later successfully campaign to have the Patriotic Society of Milan, a distinguished local cultural organization, honor Ciceri for her achievements in "country crafts." The prize, presented in 1786, served to recognize her invention of a weaving technique that used the stems of lupines to make clothes and ropes rather than the typical materials of hemp or linen.

Volta's other meaningful relationship in his single years was, unquestionably, a love affair. It nearly caused him to quit his teaching job, and perhaps jeopardized his professional future. This relationship began in December 1788, when Volta agreed to chaperone an opera singer named Marianna Paris during her visit to Como. Their romance seemingly started, and blossomed, quite quickly because Volta was already contemplating marrying her by the spring of 1789. There were complications, however, with this marriage idea, the principle one being that Marianna, as a singer, came from a lower class than did Volta. The class system was intact during this era so such a marriage would have been frowned upon—especially since Volta was a professor and universities then advocated that their students stay away from relationships with women who were in the theater.

Volta's feelings remained strong for Marianna and, in the spring of 1792, he still was seriously interesting in marrying her. He wrote about his marital ideas to Teresa Ciceri, to his archbishop brother Luigi and even to public officials. Volta's central arguments centered on the belief that marrying a deep, true love (even if the husband and wife came from different classes) was a higher good than having a loveless marriage between two people from the same class. He also asserted that marriages in which the couple came from different social strata were becoming more accepted in larger cities such as Milan. He even considered quitting his job, secretly marrying Marianna, and living off his government pension.

Volta's "modernist" marital ideas fell on deaf ears. He asked Luigi to sign a document stating that he would not oppose the marriage but could condemn it, but his brother refused to sign it. Volta even wrote to Emperor Leopold II to state his case. The Emperor actually replied several months later. He, too, would not support this nuptial.

Eventually, Volta gave up his hope of marrying Marianna, but he did not abandon her. After breaking off his relationship, Volta did persuade his brother Luigi to make a donation to her family. Luigi, however, wasn't completely certain that Volta was over the singer; he would only be convinced if his brother had a proper marriage with a more suitable bride. It did not take Volta very long to find the "proper" woman for him in Teresa Peregrini. Not only did Luigi approve of this match, but he also agreed to give Volta half of the rents from the Stampa trust to serve as a dowry.

Volta's ill-fated love affair with Marianna Paris is revealing for a couple of reasons. His attempts to go against the established class system to marry Marianna, on one hand, demonstrates that he was something of a progressive thinker and open to non-traditional ideas. The episode also reveals a bit of Volta's entrepreneurial side in the ways that he convinced Luigi to make a donation to Marianna's family and successfully negotiated for a portion of the rent money from his brother. His ability to acquire this percentage of the trust's rents wound up greatly helping him and his family, as the rental money provided a steady source of income that would balance the relative lack of compensation for his inventions, along with supplementing— and later replacing—his teaching income.

## MARRIAGE, FAMILY, AND FINAL YEARS

On September 22, 1794, Volta married Teresa Peregrini. At forty-nine years old, Volta was nearly two decades older than his bride. Teresa was the youngest daughter of Count Ludovico Peregrini, who served as the royal delegate in Como. The Count actually had been partly responsible for Volta getting his first civil servant job some twenty years earlier. Like Volta, Teresa came from a family that had some level of nobility but not much financial or social ranking.

An educated young woman, Teresa was conversant in topics related to literature and the sciences, and she spoke French and German. Volta is said to have admired Teresa's experience and intelligence. He once wrote to a friend that he chose Teresa to be his wife even though there were women who "possessed of greater physical beauty, more exalted piety, and a larger dowry" who wanted to marry him. In her own way, Teresa can be seen one of the unsung heroes in Volta's story. By virtue of providing her husband with moral and emotional support, she helped to fortify him for the inevitable frustrations and disappointments that came from working on experiments with uncertain outcomes.

Alessandro and Teresa together had three sons: Giovanni (known as Zanino) was born in 1795, Flaminio in 1796 and the youngest, Luigi, was born in 1798. All their children were born during the time of Volta's intense research on the battery. In fact, for nearly the entire first decade of his marriage, Volta spent a lot of time away from home. While his wife and children lived mainly in Como, he taught in Pavia and often traveled to Milan and across Europe for his business dealings. Because he was something of an absent husband and father in the mid-to-late 1790s, Volta must have realized that he had the tremendous opportunity to spend more time with his family in the wake of his success with the Voltaic pile when he reached his mid-fifties. Once he turned his attention more towards his family life, Volta played an extensive role in his sons' early education, teaching them basic concepts of math and grammar.

Historians also suggest that Volta and his family lived comfortably, but not especially luxuriously, for the 1800s; however, accumulating great wealth was never the prime motivation behind his work. Pancaldi, in his biography of Volta, points out that the scientist never patented his battery, nor any other invention. While patents were in use by 1800, they were not as popular as they are today, especially for natural philosophers and university professors such as Volta. Patenting battery-related devices only became common in the mid-1800s. Edison, for example, had his name on more than one thousand patents during his lifetime (1847–1931).

What Volta sought to achieve through his work was to make a difference in the world, and he certainly did so—perhaps on a grander scale than he had ever imagined. Furthermore, Volta also used his "celebrity" status as a scientist to support the generation of scientists that came after him. For instance, he used his trips to help the Austrian government acquire equipment for his university and thereby helped to improve the education of future scientists.

Volta was seriously contemplating retiring from teaching in 1804, when his old friend Napoleon paid a visit to Pavia. The Emperor was so upset upon learning about Volta's retirement plans that he was said to have exclaimed:

*If Volta's functions as a professor are fatiguing to him, let them be reduced. Let him, if he will, have to give only one lesson a year; but the University of Pavia would be struck to the heart on the day that I should permit so illustrious a name to disappear from the list of its members . . . a good general ought to die on the field of battle.*

As a result of Napoleon's entreaties, Volta remained teaching at Pavia, which must have delighted his students who idolized him as a professor.

Napoleon was one of Volta's most ardent admirers, and certainly the most powerful one. In 1805, he selected Volta to be a Knight of the Legion of Honor (the highest honor in the French Republic), and granted him a 4,000 lire annuity, which continued even after Napoleon's fall from power. The following year, he made Volta a Cavalier of the Italian Royal Order of the Iron Crown, an esteemed honor of merit that Napoleon himself had established. In 1809, Napoleon appointed Volta to be a senator of the Kingdom of Italy, a position that came with a 24,000 lire salary. He named Volta a Count of the Kingdom of Italy a year later.

Alessandro Volta's legacy was greatly boosted by Napoleon Bonaparte, and it was further enhanced by his nephew, Louis-Napoleon Bonaparte (Napoleon III). In 1801, Bonaparte created a prize to honor Volta for his achievements in inventing the Voltaic pile. Bonaparte subsequently bestowed this French Academy of Sciences award (typically referred to as the Galvanism Prize) on other scientists to salute major accomplishments in electricity. Humphry Davy, for example, was an early winner. This award was discontinued after Bonaparte fell out of power.

The award was revived in 1852 as the Volta Prize by Napoleon III, who had the same interest in science that his uncle Napoleon Bonaparte did. It is said that Napoleon III even built a version of a Voltaic pile which he then presented before the

French Academy of Sciences in the 1840s. Although presented by the Ministry of Public Instruction, the Volta Prize was funded by Napoleon III himself, and it was awarded periodically until 1888.

Among the notable winners of the Volta Prize was Alexander Graham Bell, who was received it in 1880 for his invention of the telephone. Bell would use his prize money to start several institutions, including the Volta Laboratory and later the Volta Bureau. Both made significant contributions to society: Volta Laboratory was involved in the development of such devices as the gramophone and the artificial respirator, while the Volta Bureau did important research in the areas of deafness and hard-of-hearing. The prominent work done by these two Washington D.C.-based organizations ensured that the name Volta was known into the twentieth century.

Volta's name is also attached to a relatively new science award. The Edison Volta Prize, established in 2011 by the European Physical Society, is presented biannually to honor outstanding work in the field of physics.

Although he benefited from Napoleon Bonaparte's patronage, Alessandro Volta's standing did not suffer after Napoleon abdicated as the Emperor of France and Austria returned to power in Lombardy. Unlike the professionally difficult situations Volta encountered when Austria re-took Lombardy in 1799, the 1815 political transition went smoothly. Volta was not only able to hold onto his academic post, he was promoted by Austria's Emperor Francis to the position of Director of the Faculty of Philosophy.

This new role, however, was shrouded in a bit of controversy. A few months after getting his new job, Volta wrote a declaration about his pro-religious beliefs at the request of a local canon named Giacomo Ciceri. Some observers felt Volta was forced, or at least felt strongly compelled, to publically show support for the church. During this era, it was not uncommon for prominent citizens to be questioned about their loyalties following a change in political power, and this letter could be seen as an attempt by Volta to quell lingering debates that he was somehow anti-Church.

It is important to also remember that while he did reject a life working in the Church, he still held much respect for it and its traditions. Historians also suggest that the Canon Ciceri asked Volta to write the letter because he had heard negative rumors regarding Volta, which may have been instigated by a colleague who was jealous about the recent job promotion, and aimed to get him fired.

One additional explanation for Volta's writing of this letter is that he did it as a favor. Canon Giacomo Ciceri was a young clergyman who was counseling a dying man who didn't believe in religion. Ciceri thought that the dying man would be convinced if a man in the very secular field of science stated that he believed in religion. The most prominent scientist in the area was Alessandro Volta, so it made sense that Ciceri would ask him to write this document. Further adding credence to this explanation is that Giacomo Ciceri was the son of Volta's long-time friend Teresa Ciceri, so Volta wasn't just doing this favor for just any clergyman, but one who easily could have viewed Volta as an avuncular figure.

Around this time, Volta was contacted by the University of St. Petersburg to join its illustrious faculty. Despite this unique opportunity to move to a more prestigious institution, Volta rejected the offer. He is said to have replied, "What remains for me to wish for during the few years that remain to me? To occupy myself with the education of my sons and attend a little more to my experimental research; and that is all."

This job overture, moreover, arrived a short time after his middle son, Flaminio, died at the tragically young age of eighteen. The loss of Flamino, who had shown great promise in mathematics, devastated his father. Still grieving, Volta wrote "This loss strikes me so much to heart that I do not think I shall ever have another happy day." Flaminio's death set Volta into a significant depression, which some historians have linked to his withdrawal from experimentation in this stage of his life; however he had started his transition years earlier.

During the late 1810s, Volta mainly concentrated on the education and livelihood of his two other sons. Both Giovanni and Luigi gained admission to the University of Pavia's faculty of law, and after they graduated in 1819, Volta finally retired as a professor. The family then decamped to live in the Volta country house in Camnago, a village just south of Como.

Volta did not completely isolate himself from the scientific community. In 1816, a five-volume set of his works, collected from journals and periodicals in which they had appeared, was published. The famous French physicist and mathematician Jean-Baptiste Biot described it as a valuable work, especially in exploring how Volta developed his important ideas over the course of his career.

While he generally kept himself to his family, younger scientists and dignitaries visited him to pay homage, befitting his elder statesman status as an *éminence grise* of Europe's scientific scene. Among those visiting him at Camnago were Denmark's Prince Christian and Volta's old colleague Alexander von Humboldt. The ever-modest Volta found the attention to be undeserved. "What should you want with a poor old man?" he commented rhetorically.

Sir Humphry Davy and Michael Faraday encountered this same humility when they met Volta in 1814 in Milan. Davy praised Volta's perfectly simple manners and dignity, along with his lack of pretentiousness. Munro echoed Davy's description of Volta when he wrote that "ambition, greed, or envy did not move him . . . Even his honours did not spoil him, and the habits of his boyhood still clung to him in the days of his renown."

Volta may not have been fond of all the attention directed toward him later in life. While he certainly earned the acclaim, he never let it go to his head. He remained the same man with the same values he'd always had. His apparent desire for a quieter life and sense of inner peace might have been possible because he knew that his premier invention had assured his place in history and was duly satisfied with what he had accomplished in his life.

In June of 1823, Volta suffered an apoplectic stroke from which he never fully recovered. A mild jaundice attack in the fall of 1826 further weakened his health, and on March 5, 1827, Volta succumbed to a fever that followed several days of illness. He was 82 years old. His funeral, according to reports, was a grand public event in keeping with a man of stature.

## COMO REMEMBERS ITS NATIVE SON

Shortly after Volta's death, the city leaders of Como voted to create a monument to their famous son. Due to various delays—ranging from political in-fighting to natural disasters (severe flooding as well as a cholera outbreak)—the monument was finally unveiled in 1838. At its opening ceremony, a speech was given by Francesco Mocchetti, who had been a student of

Volta and was his successor as the Chair of Physics in Como. Located in Piazza Volta, the statue bears the inscription "*A Volta—La Patria*" ("*To Volta—The Nation*").

In 1831, the Volta family built a tomb for him in the Camnago cemetery. Shaped like a small temple, it was decorated with allegorical bas-reliefs representing several of his inventions: the eudiometer, the flammable air gun, the electrophorus, the condenser, and the Voltaic pile.

The University of Pavia similarly erected several sculptures of Volta that recognized his contributions to science and his affiliation with the school. One statue was done in 1878 to commemorate the centennial of Volta being named the chair of the university's physics department. The statue's inauguration was such a major event that Italian Prime Minister Benedetto Cairoli attended. The university also founded The Institute of Experimental Physics in Volta's honor in 1880. It is still in existence, though now under the name "The Department of Physics A. Volta."

On a rather macabre note, the University of Pavia received permission from the Volta family in 1875 to allow two anthropologists, Cesare Lombroso and Paulo Mantegazza, to disinter Volta's remains and perform a craniometrical examination. The phrenological findings were not revelatory. Volta's skull was similar in size to those of ancient Romans, while his brain weight was just over fifteen ounces more than the typical weight. Lombroso's paper reportedly provoked laughter when it was read at the Istituto Lombardo Accadademia di Scienze e Lettere (Lombard Institute Academy of Science and Letters). This reaction might be attributed to the fact that Volta not only was a founding member of this Napoleon-established Institute, but its first chairperson too.

In 1841, Florence constructed a tribute to great Italian

scientists and their experimental spirit on the first floor of the Science Museum of La Specola. Called "Galileo's Tribune," it showcased the crucial role that Italians have played in the story of science. Its centerpiece is a set of seven frescos highlighting important moments in Italian scientific history. The seventh and culminating fresco depicts Volta showing his battery device to Napoleon.

Volta also serves as a primary figure in another notable fresco, "The Triumph of Science," that Nicolò Barabino painted in the Palazzo Orsini in 1876. While populated by such historical figures as Columbus, Watt, Newton, Galileo, and Gutenberg, it is Volta who stands as the painting's central focus. In this portrayal, Volta is demonstrating his Voltaic pile.

Volta received a quite unique type of posthumous recognition, courtesy of the French philosopher Auguste Comte. In 1848, the father of positivism proposed a new calendar, aptly entitled the Positivist Calendar. The months were to be themed around different concepts, and feature names of real people associated with these themes (i.e. ancient philosophy and Aristotle; modern drama and Shakespeare). While Comte did not seem to regard electricity as an evolved branch of physics, he still described Volta as "immortal" and featured him in the month devoted to modern science along with Newton, Galileo, Copernicus, and Kepler—as well as Volta's contemporaries Priestley, Lagrande, and Lavoisier.

For many years, visitors could go into the apartments where Volta once had lived in Como and view his books, papers, machines, inventions, and even his cane and tobacco pouch. In the early 1860s, however, the Volta family found it necessary to sell their collection of Volta's effects. Fortunately, the Lombard Institute

Academy of Science and Letters stepped in to purchase this collection for 100,000 lire before it could be sold off and dispersed. Volta's historic items were then housed by the academy at the Palazzo Brera as the "Cimeli di Volta" (memorabilia of Volta).

To honor the centennial anniversary of Volta inventing the electric battery, the city of Como decided to host a grand exhibition entitled the "International Electrical Exposition." Libraries, museums, and universities searched their archives for items, particularly material and apparatuses used by Volta, to contribute to this event. An exhibition hall was built especially for the occasion, and King Umberto of Italy presided over the opening festivities on May 20, 1899.

Unfortunately, the celebratory atmosphere was shattered on July 8 when a fire broke out and burned down the hall. The vast majority of the Volta relics—many of them irreplaceable—were destroyed in the fire, which was caused, ironically, by a faulty electrical wire. Happily, a few items did survive, such as the Senatorial sword that Napoleon had given Volta. In a fortuitous stroke of luck, the collection of Volta memorabilia housed at the Palazzo Brera had not been sent to the exposition, so it remained intact.

## THE NAMING OF THE VOLT

The most enduring and universally known honor for Volta came in 1881 at the International Electrical Congress. This Paris-based convention, also known as the International Congress of Electricians, included 250 delegates from twenty-eight countries. A main purpose behind this gathering was to devise standardized measurements for common use. For years, many concepts had different terms, which understandably caused confusion and hampered the ability to make precise, and universally accepted, calculations.

After plenty of haggling and much backroom politicking, several units of measurements were named, with the unit of electric potential named the *volt* after Alessandro Volta. The Congress also named a unit of electrical capacitance the *farad* for Michael Faraday; a unit of current the *ampère* for André-Marie Ampère; a unit of electrical resistance the *ohm* for German Georg Simon Ohm and a unit of electric charge the *coulumb* for Charles-Augustin de Coulomb.

It should be noted that Volta was the only Italian scientist whose name was selected for one of these terms. Helping Volta's cause, however, was the backing that he apparently had from the French delegates, which came about because of the well-known support that Napoleon had shown Volta.

The *volt* (and its symbol: V) is one of the International System of Units that represents, according to the Oxford dictionary, "the difference of potential that would carry one ampere of current against one ohm resistance." Volta undoubtedly would have been thrilled with having this particular term named after him since he was quite interested in standardizing measurements and equipment. Furthermore, *volt*, which reflects the energy that can be released if an electrical current is free to flow, also well represents Volta the man, since his whole life was about unlocking human and scientific potential.

The word *volt*, in a very real sense, has made Volta truly immortal since it and its variants have become deeply embedded in our thought and language. On the scientific front, it forms the root for names of such instruments as the *voltmeter*, which calculates the electrical potential difference in an electric circuit between two points, and the *voltameter*, which measures the quantity of an electric charge through electrolytic action. The term *voltage* not only means electrical potential difference, but also refers to energy, action, excitement, and enthusiasm. Moreover, use of the word *volt*, as will be examined shortly, has expanded in the modern era. It should also be noted that Volta's old colleague and rival, Luigi Galvani, has been immortalized in words such as *galvanism*, *galvanic* and *galvanize*, to name a few.

## VOLTA IN THE TWENTIETH CENTURY AND BEYOND

Even after its ill-fated 1899 exhibition, the city of Como has continued to shine a light on its favorite son. To commemorate the centennial of his death, Como staged The Voltain Exposition in 1927. At a cost of more than seven million lire at that time, this enormous event became more of a civic event to promote Como than merely an Alessandro Volta celebration. The Exposition featured large exhibitions on telegraphy (telegraph pioneer Guglielmo Marconi was an honorary Exposition president, along with Benito Mussolini) and silk (a traditional Lombardy industry), as well as dog shows, musical competitions, and horticultural and outdoor sports exhibits. Since the city didn't have a stadium, Sinigaglia Stadium was constructed, which still stands today.

Science was not entirely ignored at the Voltian Exposition; it did present the International Conference on Physics, which attracted sixty-one scientists from fourteen countries. An impressive roster of attendees included twelve Nobel Prize winners (such as Niels Bohr, Max Planck, Hendrik Lorenz, and Marconi) as well as three future Nobel winners (Max Born, Enrico Fermi, and Werner Heisenberg). In fact, it stands one of the first post-World War I international science conference that German and Austrian scientists were able to attend.

Despite boasting three generations of top physicists, this conference did not result in significant contributions to the science world. Its most notable presentation was one by Bohr on quantum physics, but his speech did not cover original information. The lack of interesting news emanating from this convention sadly mirrors the organizers' lack of interest. If not for a last-minute bank loan, the conference would have been a casualty of the Exposition's severe money shortfall.

Considering its epic size and its Mussolini connection, the Exposition would seem to have been part of that leader's propaganda campaign to encourage Italian pride. But Mussolini, as it turns out, had little direct involvement in the event. In fact, the prime minister of Italy (1927 was five years into the Fascist rule and before Mussolini declared himself dictator) canceled a speech scheduled for the Exposition and rejected a request for emergency money from its short-funded organizers.

But the Exposition did leave two lasting legacies. One is the Tempio Voltiano or Volta Temple and the other is the Voltiano Lighthouse. Built in honor of Volta, the lighthouse is electrically lit, and nightly shines beams in the Italian national colors of green, white, and red. The Temple, which was inaugurated in 1928, stands as both a museum and a monument to Volta. Housing a large collection of his devices, instruments and papers, this permanent memorial serves as a repository of the Volta artifacts that survived the 1899 fire, as well as items that have been collected since then. The building features sixteen showcases displaying Volta's life and his important work in the realm of science.

Como has continued to celebrate milestone anniversaries

of Volta's life. His birth was honored on its 200th anniversary (1945), the 250th (1995), and the 150th anniversary of his death was saluted in 1977. The city also hosted events that recognized the 150th anniversary of the Voltaic pile's invention (1949) as well as its bicentennial in 1999.

Among his hometown's most recent tributes, one took place in October 2015 when the city of Como unveiled a stunning sculpture entitled "Life Electric" that was designed by the award-winning architect and artist Daniel Libeskind. Working with the Amici di Como, Como's business association, Studio Libeskind created a 54-foot high public sculpture that stands tall in Como's waterfront. Studio Libeskind describes this modernist stainless steel and stone work, which becomes an illuminated beacon at night, as "representing the two poles of energy—a nod to Volta's invention of the battery."

And in 2000, the University of Pavia opened a new college named Collegio Alessandro Volta after their famous former professor. The school's mission involves accepting a large number of international students as well as presenting an academic program that explores public understanding of science and the impact of science on society. These two principles seem particularly fitting for a college named for Volta, since his life and work were very much international in scope and his primary interest was how his scientific inventions could benefit humanity.

The nation of Italy has also saluted Volta's legacy over the years. From 1984 to 2002, the country's 10,000 lire note featured an image of Volta on the front, along with the Voltaic battery and the winged lion of St. Mark (the symbol of Venice) above three shields of Pisa, Genoa and Amalfi. The back of the banknote depicts Como's Tempio Voltiano museum, and its watermark was the "Portrait of Alessandro Volta."

Before the euro became the legal tender in Italy, the popular 10,000 lire banknote (worth about $6 or $7) represented a mass-circulation currency that almost every Italian person commonly used during that era. Between September 3, 1984 (its issue date) and February 28, 2002 (when it was withdrawn from circulation), the Bank of Italy issued 3.2 billion notes featuring Volta.

It is natural that Volta has been honored in Italy, but he has been recognized in America as well. In northwestern Washington D.C., Volta Park exists on Volta Place. The two spots bear Volta's name because the Volta Laboratory and Volta Bureau were located in that neighborhood. Volta Park actually has its own historical significance dating back to Volta's time. From 1802–1909, it was the site of the Presbyterian Burying Grounds, which was arguably Washington D.C.'s most prominent cemetery in the first half of the nineteenth century.

Across the United States, several schools have been named for Volta. Alessandro Volta Elementary School is part of the Chicago public school system. Located in the multi-ethnic Albany Park neighborhood, the school's motto, "Where the World Learns Together," reflects a sense of global thinking that would have been shared by its namesake. The central California city of Los Banos is home to an elementary school named for Alessandro Volta too. The town is located on the southern side of the Silicon Valley, which is fitting considering that the existence of the tech industry can be traced back to Volta. Perhaps the American school whose name is more directly inspired by the legendary Lombard electrician is the Volta Line School. Based in northeastern Oregon, this nonprofit trade school trains adult students for work in the electrical construction industry—a type of work that would not exist were it not for Volta.

Probably the best-known commercial and cultural reference to Volta in the twenty-first century is the electric car. When Chevrolet debuted its electric hybrid car to America in 2011, the company quite appropriately named it the Volt. Years later, *Car and Driver* described the initial edition of this vehicle as being "revolutionary," and proclaimed that it "has remained the gold standard for plug-in hybrid vehicles."

Unless you are an auto enthusiast, you might not know what there also was an electric car named for Volta prior to the Volt. In 2004, Toyota debuted the prototype for a stunning coupe called the Alessandro Volta. Manufactured by Toyota and designed by Italdesign-Giugiaro, this sports car (a two-door coupe) featured a hybrid electric engine. Described as "tragically cool," it could accelerate from zero to sixty in four seconds, and reach a top speed of 155 mph. This innovatively designed vehicle debuted at the 2004 Geneva Motor Show, but the concept car never went into production because of the high expense of its mass manufacture.

Volta's name is prominent in the motorcycle and motor-bike world also. Spain-based Volta Motorbikes introduced its first electric motorcycle quite appropriately at the 2012 Milan Motorcycle Trade Show. Milan also is home to Italian Volt, which produces customizable electric motorcycles utilizing 3D printing. Both firms salute their namesake through the use of electric engines and their innovative use of technology.

Volta's name has been used for a variety of products and companies. On the high-tech side, Volta is the name that the computer company NVIDIA selected for its mega-powerful GPU that contains twenty-one billion transistors used in the world's largest super-computers, such as the major cloud service providers and data center service manufacturers. Perhaps not coincidentally, the Volta (aka V100) is part of NVIDIA's Tesla line of Data Center GPUs.

On the other end of the consumer spectrum is the Mountain Dew Voltage beverage, a sweet, carbonated concoction that combines raspberry and citrus flavors with ginseng. Its name was selected to suggest the rush of energy experienced when drinking it.

Cirque du Soleil entitled one of their stage spectaculars, *VOLTA*. The pioneering, award-winning theatrical troupe explains this show as "about being true to oneself, fulfilling one's true potential, and recognizing one's own power to make it possible." This description certainly reflects the philosophy that the man himself lived by.

Efforts to honor and recognize Volta have even extended to outer space. Among the 19,000-plus named minor planets, one has been named Volta—8208 that was discovered by the astronomers Piero Sicoli and Pierangelo Ghezzi in 1995. The moon, moreover, is home to the Volta Crater, which is located on a lunar crater near the northwest limb of the moon. There also is a Galvani Crater on the moon, and, just as Volta's reputation overshadows Galvani's on the earth, so does the larger Volta lunar crater overshadow the nearby Galvani crater.

In 2015, the 270th anniversary of Alessandro Volta's birthday was celebrated with large and small events. The Solomon Islands, a South Pacific nation comprising hundreds of tiny islands, issued a four-part series of stamps to commemorate Alessandro Volta that year. Set against a beautiful backdrop of Volta Hall at the University of Pavia, the four-stamp sheet contains (in clockwise order) a stamp of Volta himself with his Voltaic pile; one show-ing Volta explaining "the principle of the 'electric column' to Napoleon in 1801;" one depicting Luigi Galvani that references how the Galvani-Volta debate led to the electric battery; and the last featuring a statue of Alessandro Volta in Como, Italy, and the Tempio Voltiano. These stamps reflect the continuing inter-est in Volta's life and work around in the world.

Volta's 270th birthday was also acknowledged on a global scale by Google, when the search-engine giant created a special Google Doodle of Volta on February 18, 2015. Mark Holmes, the artist behind the Doodle, commented at that time that the design "was particularly thrilling given he was the eighteenth-century Italian physicist, chemist, and electrical pioneer who invented the first electrical battery." Wishing to connect Volta's important inventions to the man whose face isn't well known today, Holmes came up with artwork that resembles an old-time advertisement for the electric battery, with a Voltaic pile, and the letters in Google light up to represent the concept of increased voltage.

Besides showcasing Volta's continuing relevance to our lives, the Google Doodle led to a flurry of attention, providing the inventor some welcome posthumous recognition. Stephanie Mlot really drove home this message in a *PC Mag* article in which she stated that "without Alessandro Volta, there'd be no portable DVD players, power tools, pacemakers, or Energizer Bunny."

## FINAL THOUGHTS

It is rather amusing to see the many different ways in which Alessandro Volta's name shows up in the twenty-first century—a sugary soft drink, an electric car, a theatrical extravaganza, and a crater on the moon. What is staggering is to consider the myriad applications that Volta's invention of the battery has in our lives today. We get our entertainment from the games and toys powered by batteries. We communicate and work through the computers, phones, and other devices that are powered by batteries. We build homes with power tools and save lives through the battery-powered medical equipment. The list continues ad infinitum.

It goes without saying that our society owes a huge thanks to Alessandro Volta. He unquestionably resides in the pantheon of great inventors in the history of science and technology. While he might not be a household name, or as well known as Alexander Graham Bell or Albert Einstein, Volta is not a forgotten man. He has been the subject of 1,500 books and papers, has been included in virtually every important guide to inventions and profiled on a multitude of websites that discuss famous scientists.

It is certainly astonishing too to consider how the world has changed culturally, economically and technologically since Alessandro Volta was born more than 270 years ago (some three decades before the founding of the United States, to put it in an historical perspective), and the part that he played in those changes.

It is also inspiring to consider how Alessandro Volta lived his life during his own age. Very much a self-made man, Volta was persistent and relentless in pursuing his goals. He knew how to network before the term was invented. He realized how to build upon his own work and ideas, and that of others, to create new inventions and ideas, and push the world forward. Not only did he possess an imaginative mind, Volta also had the talent to turn his ideas into practical devices. He was a thinker and a maker. A dreamer and a doer.

# Acknowledgments

This book is greatly indebted to the work done by Giuliano Pancaldi in his definitive Alessandro Volta biography, *Volta: Science and Culture in the Age of Enlightenment* (Princeton University Press, 2003), as well as Bern Dibner's earlier biography *Alessandro Volta and the Electric Battery* (Franklin Watts, 1964). Both works are highly recommended for those wanting to learn more about this great Italian scientist.

Besides Pancaldi's and Dibner's books, a number of other sources were utilized in the writing of this book. Robert E. Schofield's biography, *The Enlightened Joseph Priestley: A Study of His Life and Work from 1773 to 1804* (Pennsylvania State University Press, 2004), and *The First Scientific American: Benjamin Franklin and the Pursuit of Genius* (Basic Books, 2006) by Joyce E. Chaplin were sources for Chapter One, as was Thomas Sowell's article "The Einstein Syndrome" (August 31, 2001).

A profile on Volta in the July 22, 1899 issue of *Scientific American* Vol. 80–81 and the Volta biographical profiles on the Electronics Notes website and the Famous Scientists website were used as reference sources in Chapter One and several subsequent chapters too.

Sources for Chapter Two's historical background include the books *Science For A Polite Society: Gender, Culture, And The Demonstration Of Enlightenment* by Geoffrey V. Sutton (Avalon Publishing, 1997); *Fire and Light: How the Enlightenment Transformed Our World* by James MacGregor Burns (Thomas Dunne, St. Martin's Press, 2013); *A History of Science, Volume 2* edited by Henry Smith Williams (Harper, 1904); and *1001 Inventions That Changed the World* edited by Jack Challoner (Barron's Educational Series, 2009), which also was a source for Chapter Ten.

Helpful too for Chapter Two were David Herres' article "Francis Hauksbee and Static Electricity Generation" (Test & Measurement Tips, January 23, 2015); Matthew White's article "The Rise of Cities in the 18th Century" found on The British Library website; the "Electricity" chapter by Christine Blondel in *Encyclopedia of the Enlightenment* edited by Michel Delon (Routledge, Taylor & Francis Group, 1997); and "Learned Societies" by Michael Heffernan in *The SAGE Handbook of Geographical Knowledge*, edited by John A. Agnew, David N. Livingstone (SAGE Publications, 2011), along with James E. McClellan's contributions in *Encyclopedia of the Enlightenment*, ed. Alan Charles Kors (Oxford: Oxford University Press, 2003) and *The Cambridge History of Science. Vol. 4.*, edited by Roy Porter (Cambridge University Press. 2003).

Additional background information for Chapter Two was found in an article entitled "Enlightenment" on History.com; "The Electric 'Circuit'" from SparkMuseum.com; Laura J. Snider's entry of William Whewell in Stanford Encyclopedia of Philosophy; Palmira Fontes da Costa's chapter, "Women and the Popularization of Botany in Early Nineteenth-Century

Portugal: The Marquise of Alorna's *Botanical Recreations*," in *Popularizing Science and Technology in the European Periphery, 1800–2000* edited by Dr. Agustí Nieto-Galan, Dr. Enrique Perdiguero, and Dr. Faidra Papanelopoulou (Routledge, Taylor & Francis Group 2009); *The Britannica Guide To The Ideas That Made The Modern World: The people, philosophy and history of the Enlightenment* (Encyclopedia Britannica 2008); Loren Butler Feffer's entry on "The Emergence of Scientific Societies" found on encyclopedia.com; "The Electric 'Circuit'" from SparkMuseum.com; "Nollet Electrifies Royal Guard" from Engineering and Technology History Wiki; and Ralph S. Wolfe's "Pistola di Volta," (*ASM News* 70 2004), which was utilized in Chapter Four too.

The sources for Chapter Two's discussion of eighteenth-century publishing were "The History of Print from 1700 to 1749" from Prepressure.com; "The History of Magazines" from Magazines.com; (https://www.magazines.com/history-of-magazines); "The Book: 1450 to the Present" from Eduscapes.com; Max Roser's article "Books" from OurWorldInData.org, and Philip Soundy Unwin, George Unwin and David H. Tucker's "History of Publishing" entry in Brittanica.com. While "Travel as Education in 16th–19th Century Europe: From Grand Tours to Working Men's Excursions" from WhyTravel.org and "Brief History of Railroads in Europe" and "Railroads & the Development of the Idea of Europe" from EuropeanRailroads.edu provided valuable information on 18th century transportation.

Chapter Three's sources include Joseph S. Nye, Jr. *Soft Power: The Means to Success in World Politics* (PublicAffairs, 2005); the essay "European Powers in the 18th Century: Alliances, Wars

& the Balance of Power" from Study.com; "The Guardian View on the Hanoverian Monarchy" from TheGuardian.com; and Britannica.com's entry "18th-century Britain, 1714–1815."

*The Cultural Origins of the French Revolution* by Roger Chartier (Duke University Press, 1991) also provided Chapter Three background on the Enlightenment as did Robert Wokler's thorough "Enlightenment" entry in Encyclopedia. com. Information on Prince Dmitri Gallitzin was gathered from Peter De Clercq's chapter "The Instruments of Science: The Market and the Makers" in *The History of Science in the Netherlands: Survey, Themes and Reference*, edited by Klaas Van Berkel, Albert Van Helden, L. C. Palm (BRILL, 1999) and Lissa Roberts' chapter "Going Dutch: Situating Science in the Dutch Enlightenment" in *The Sciences in Enlightened Europe*, edited by William Clark, Jan Golinski, Simon Schaffer (University of Chicago Press).

Thomas B. Greenslade Jr.'s section on "Volta's Pistol" found in *Instruments for Natural Philosophy* from Kenyon.edu aided both this chapter and Chapter Six, while Lucio Fregonese's profile "Volta, Alessandro Giuseppe Antonio Anastasio" in Encyclopedia.com was useful for this chapter, and for Chapters Six and Eleven too.

Ralph S. Wolfe's "An Historical Overview of Methanogenesis" chapter in *Methanogenesis: Ecology, Physiology, Biochemistry & Genetics* edited by James G. Ferry (Chapman & Hall, 1993) and Richard Leroy Myers's *The 100 Most Important Chemical Compounds: A Reference Guide* (ABC-CLIO, 2007) aided in Chapter Four's discussion of methane as did "Early Investigations of Methane in New Jersey" from Montclair.edu, "Earliest Account of Marsh Gas" from *The Writings of Benjamin*

*Franklin, Vol. 6, 1773-1776* (New York: Macmillan, 1906), and Linda Dailey Paulson's October 26, 2015, article entitled "What Is Methane?" *Electricity and Magnetism: A Historical Perspective* by Brian Scott Baigrie (Greenwood Press, 2007), meanwhile, proved helpful in the discussion of Franklin's "One-Fluid" theory.

The examination of Volta's interest in meteorology in Chapter Five was added by the books *Amedeo Avogadro: A Scientific Biography* by Mario Morselli (Springer, 1984; reprint, 2012) and Peter Moore's *The Weather Experiment: The Pioneers Who Sought to See the Future* (Farrar, Straus and Giroux, 2015) as well as Marco Ciardi's chapter entitled "Falling Stars, Instruments and Myths: Volta and the Birth of Modern Meteorology," *Nuova Voltiana: Studies on Volta and His Times, Vol. 3* edited by Fabio Bevilacqua and Lucio Fregonese (Hoepli, 2001), and Jim Burton's "Meteorology" chapter in *Britain in the Hanoverian Age, 1714–1837: An Encyclopedia*, edited by Gerald G. Newman and Leslie Ellen Brown (New York: Garland, 1997). Theodore S. Feldman's chapter "Meteorology," in the John L. Heilbron-edited *The Oxford Companion to the History of Modern Science* (Oxford University Press, 2003) and Feldman's chapter on "Late Enlightenment Meteorology" from *The Quantifying Spirit in the 18th Century*, edited by Tore Frängsmyr, J. L. Heilbron, and Robin E. Rider (Berkeley: University of California Press, 1990) also aided in the writing of Chapter Five.

The historical background sources for Chapter Six include *Smithsonian Timelines of History* (DK Publishing, 2011), Trevor H. Levere's Britannica.com profile of Henry Cavendish, and the Antoine Lavoisier biography at FamousScientists.org, along with William Bynum's *A Little History of Science* (Yale University Press, 2012) and the "Sketch of Alessandro Volta," article from an 1892 edition of *Popular Science* from TodayInSci.com, which served as a source for Chapter Eleven too.

The discussion of the Volta-Galvani rivalry in Chapter Seven was aided by Marco Piccolino and Marco Bresadola's *Shocking Frogs: Galvani, Volta, and the Electric Origins of Neuroscience* (translated by Nicholas Wade; Oxford University Press, 2013); Stanley Finger's *Minds Behind the Brain: A History of the Pioneers and Their Discoveries* (Oxford University Press, 2000); Alex Boese's *Elephants on Acid: And Other Bizarre Experiments* (Mariner Books, 2007), and Marcello Pera's *The Ambiguous Frog: The Galvani-Volta Controversy on Animal Electricity* (Princeton Legacy Library, 2014).

Additional background material came from Dante D'Epiro's work "Thirty-nine: Trailblazers in Electricity: Galvani and Volta," in *Sprezzatura: 50 Ways Italian Genius Shaped the World*, edited by Peter D'Epiro and Mary Desmond Pinkowish (Anchor Books, 2001); John Langone's "Mind and Behavior," in John Langone, Bruce Stutz, and Andrea Gianopoulos's book *Theories for Everything: An Illustrated History of Science from the Invention of Numbers to String Theory* (National Geographic, 2006); Bern Debner's chapter on "Luigi Galvani" in Brittanica. com; a Galvani profile by Matthias Tomczak from his website, and the *Neuroscientifically Challenged* "History of Neuroscience: Luigi Galvani" article.

*Engineering in History* by Richard Shelton Kirby, Sidney Withington, Arthur Burr Darling, and Frederick Gridley Kilgour (Dover Publications, 1990; originally published by the McGraw-Hill Book Company, 1956) was valuable for Chapter Eight as was Anand Kumar Sethi's *The European Edisons: Volta,*

*Tesla, and Tigerstedt* (Palgrave Macmillan US, 2016), which also proved helpful for Chapter Eleven.

Giulo Martinez's August 26, 1899, article "The Volta Centenary" from *The Electrical World and Engineer, Volume 34*, and Will and Ariel Durant's *The Age of Napoleon: A History of European Civilization from 1789 to 1815* (Simon and Schuster, 1975) were sources for Chapter Nine. Additional historical details on Napoleon Bonaparte were gathered at the History. com Bonaparte bio; the "History of the Austrian Empire" found at HistoryWorld.net; the "Treaty of Amiens" Britannica. com entry and the historical timeline created on the Fondation Napoléon website Napoléon.org.

Two John L. Heilbron books, *Electricity in the Seventeenth and Eighteenth Centuries* (University of California Press, 1979) and *Physics: A Short History from Quintessence to Quarks* (Oxford University Press, 2015), provided source material for Chapter Ten. Other books referenced for Chapter Ten are Jürgen Osterhammel's *The Transformation of the World: A Global History of the Nineteenth Century* (translated by Patrick Camiller; Princeton University Press, 2014); Richard Holmes' *The Age of Wonder: How the Romantic Generation Discovered the Beauty and Terror of Science* (Pantheon Books, 2008); Ernest Freeberg's *The Age of Edison: Electric Light and the Invention of Modern America* (Penguin Press, 2013); Henry Schlesinger's *The Battery: How Portable Power Sparked A Technological Revolution* (Smithsonian Books/HarperCollins, 2010); Jolyon Goddard's *National Geographic Concise History of Science and Invention: An Illustrated Time Line* (National Geographic, 2010); *Popular Mechanics Gadget Planet: 150 Gizmos & Inventions That Changed the World* (Hearst Books, 2014), and *Science: The Definitive*

*Visual Guide*, edited by Adam Hart-Davis, (DK Publishing, 2009).

The sources for the timeline included the Britannica.com entries on Pavel Yablochov; the Alexander Graham Bell profile which was written by David Hochfelder, and the Thomas Alva Edison entry which was penned by Matthew Josephson and Robert E. Conot, along with the entry on the Arc Lamp and Piezoelectricity.

Additional timeline information was found in "Lamp Inventors 1880-1940: Carbon Filament Incandescent"; the Ernst Werner Von Siemens profile at FamousScientists.org; the *Daily Mail* article "Clearly the Work of a Bright Spark: Victorian House That Was First In World To Be Powered By Water Will Be Lit Up Using Hydroelectricity Again" by Mark Duell (July 29, 2014), and the Wired.com pieces: "Let There Be Light—Electric Light" by Priya Ganapati and "The Birth of the Microphone" by Matthew A. Shechmeister Jan. 11, 2011).

Chapter Ten sources also included Fabio Bevilacqua's book review "The Investigations and Inventions of Volta," from the November-December 2003 *American Scientist*; J.B. Shank's Britannica.com profile of André-Marie Ampère; Sean Trainor's Time.com piece "History: What the Digital Age Owes To The Inventor of Morse Code," April 27, 2016, and Franco Decker's January, 2005 essay "Volta And The Pile" from Electrochem. org.

Information on batteries were gleaned from Michelle Z. Donahue's *Smithsonian Insider* article "Unplugged: 5 Batteries That Gave The World A Jolt" that came out on May 15, 2015; Isidor Buchmann's article "When Was the Battery Invented?" from the National Association of Amateur Radio's website aarl. org; Ed Yong's December 13, 2017, Atlantic.com piece "A

New Kind of Soft Battery, Inspired by the Electric Eel" and the April 2, 2018, article "What's Current With Your Battery" article from Electric Bike Action Magazine.

Contributing to the discussion of batteries too were Sean Duke's September 15, 2015, essay "Battery Technology Has Changed Little Since Volta, But Are Longer Lasting Batteries Finally Here?" from ScienceSpinning.com; Dom Galeon's article "The Next Electric Car to Make it to Market May Have a One Minute Charge Time" from Futurism.com; "How Batteries Changed The World," published September 6, 2017, TonikEnergy.com; and "5 Fast Facts about Hydrogen and Fuel Cells" from Energy.gov, October 4, 2017; as well as the conference paper "Lithium-Ion Rechargeable Batteries on Mars Rover" by B.V. Ratnakumar, M.C. Smart, R.C. Ewell, L.D. Whitcanack, K.B. Chin, and S. Surampudi, that was presented Aug. 15, 2004, at the 2nd International Energy Conversion Engineering Conference.

John Munro's *Pioneers of Electricity Or, Short Lives of the Great Electricians* (Religious Tract Society, 1890), Paul Parsons's *Science in 100 Key Breakthroughs* (Firefly Books, 2011), and Seth Fletcher's *Bottled Lightning: Superbatteries, Electric Cars, and the New Lithium Economy* (Hill and Wang/Farrar, Straus and Giroux, 2011) furthermore served as sources for both Chapter Ten and Chapter Eleven.

Nancy Forbes and Basil Mahon's *Faraday, Maxwell, and the Electromagnetic Field: How Two Men Revolutionized Physics* (Prometheus Books, 2014), Lutz D. Schmadel's *Dictionary of Minor Planet Names* (Springer, 2003), and Karl Alois Kneller's *Christianity and the Leaders of Modern Science; A Contribution to the History of Culture in the Nineteenth Century* (B. Herder, 1911 retrieved from Francesblogg.com), provided supporting material for Chapter Eleven as did Chuck Layne's profile on "Alessandro Giuseppe Antonio Anastasio Volta" and Jürgen Teichmann's article "Georg Christoph Lichtenberg: Experimental Physics from the Spirit of Aphorism."

Additional Chapter Eleven sources include J.J. Fahie's article "The Institution's Visit To Como and Milan," from the March 6, 1903, edition of the *Supplement to The Electrical Engineer: An Illustrated Record and Review of Electrical Progress, Volume 31*; Gérard Borvon's September 10, 2012, essay "History of Electricity: The History of Electrical Units"; the profile on Alessandro Volta found on the Alessandro Volta Foundation's website, alessandrovolta.it; *The Man of Genius* by Cesare Lombroso, and Graziano Magrini's Tribuna di Galileo article "Scientific Itineraries in Tuscany," translated by Catherine Frost.

Useful to Chapter Eleven's discussions of Alexander Graham Bell and Volta Park were the National Park Service's "Washington, DC A National Register of Historic Places Travel Itinerary"; Volta Park's History on the Friends of Volta Park's website; "Mapping the Hidden Locations of 10 Long-Gone Burial Grounds" by Michelle Goldchain; The Georgetown Metropolitan's "What Would Volta Park Be Without Napoleon Bonaparte" article; and Candace Wheeler's October 2, 2012, Washington Post article "Construction Boon in D.C. Leads to Discoveries of Old Burial Sites."

The Academic Dictionaries and Encyclopedias, meanwhile, was helpful for background information on the Volta Prize, while The European Physical Society was a good place to learn about the ESP Edison Volta Prize.

Valuable Chapter Eleven background material was provided by Barry Lillie's September 7, 2014, *Italy Magazine* article "Tempio Voltiano on Lake Como: A Museum for Alessandro

Volta"; the Como Lake Experience website; the City of Como website; the University of Pavia website; and the Studio Libeskind website.

Details on Alessandro Volta Elementary School were found on the school website, while information on the Chevrolet Volt, Toyota Volta, and the Volta Motorbike came from Car and Driver, Hybridcars.com, and Electricmotorcyces.nets, respectively.

Mark Holmes' essay on Google.com and Stephanie Mlot's "Electric Battery Creator Alessandro Volta Honored With Google Doodle" from the February 18, 2015 *PC Mag* had information about the Volta Google Doodle, while details on the Solomon Islands' Volta stamps was found on the official Solomon Islands stamp agency.

The sites etymonline.com, definitions.net, oxforddictionaries.com, and banknotes.com were utilized to gather technical and historical details for the book.

## About the Author

Michael Berick is a writer and journalist whose work has appeared in outlets such as the *Los Angeles Times*, *Entertainment Weekly*, *LA Weekly*, *AAA Westways Magazine*, and the *San Francisco Chronicle*. He has written about European chocolate destinations, reviewed artist Ed Ruscha's retrospective, and penned press material for the Grammy-nominated boxset, *Battleground Korea: Songs And Sounds Of America's Forgotten War*. He also might possibly be the only music critic to have voted in both the Fids and Kamily Music Awards and the *Village Voice*'s annual Pazz & Jop Poll. Hailing from Cleveland, Ohio, Berick currently lives in Los Angeles with his wife—playwright/screenwriter Jennifer Maisel—and their daughter and dog.

# NOW AVAILABLE FROM THE MENTORIS PROJECT

*America's Forgotten Founding Father*
*A Novel Based on the Life of Filippo Mazzei*
by Rosanne Welch, PhD

*A. P. Giannini—The People's Banker*
by Francesca Valente

*The Architect Who Changed Our World*
*A Novel Based on the Life of Andrea Palladio*
by Pamela Winfrey

*A Boxing Trainer's Journey*
*A Novel Based on the Life of Angelo Dundee*
by Jonathan Brown

*Breaking Barriers*
*A Novel Based on the Life of Laura Bassi*
by Jule Selbo

*Building Heaven's Ceiling*
*A Novel Based on the Life of Filippo Brunelleschi*
by Joe Cline

*Building Wealth*
*From Shoeshine Boy to Real Estate Magnate*
by Robert Barbera

*Building Wealth 101*
*How to Make Your Money Work for You*
by Robert Barbera

*Christopher Columbus: His Life and Discoveries*
by Mario Di Giovanni

*Dark Labyrinth*
*A Novel Based on the Life of Galileo Galilei*
by Peter David Myers

*Defying Danger*
*A Novel Based on the Life of Father Matteo Ricci*
by Nicole Gregory

*The Divine Proportions of Luca Pacioli*
*A Novel Based on the Life of Luca Pacioli*
by W.A.W. Parker

*Dreams of Discovery*
*A Novel Based on the Life of the Explorer John Cabot*
by Jule Selbo

*The Faithful*
*A Novel Based on the Life of Giuseppe Verdi*
by Collin Mitchell

*Fermi's Gifts*
*A Novel Based on the Life of Enrico Fermi*
by Kate Fuglei

# FUTURE TITLES FROM THE MENTORIS PROJECT

*A Biography about Rita Levi-Montalcini*
and
Novels Based on the Lives of:
*Amerigo Vespucci*
*Andrea Doria*
*Antonin Scalia*
*Antonio Meucci*
*Buzzie Bavasi*
*Cesare Beccaria*
*Father Eusebio Francisco Kino*
*Federico Fellini*
*Frank Capra*
*Guido d'Arezzo*
*Harry Warren*
*Leonardo Fibonacci*
*Maria Gaetana Agnesi*
*Mario Andretti*
*Peter Rodino*
*Pietro Belluschi*
*Saint Augustine of Hippo*
*Saint Francis of Assisi*
*Vince Lombardi*

For more information on these titles and
the Mentoris Project, please visit
www.mentorisproject.org

Printed in Great Britain
by Amazon

59496609R00118